The 1992/93 academic year at the Mathematical Sciences Research Institute was devoted to complex algebraic geometry. This volume collects survey articles that arose from this event, which took place at a time when algebraic geometry was undergoing a major change. To put it succinctly, algebraic geometry has opened up to ideas and connections from other fields that have traditionally been far away.

The editors of the volume, Herbert Clemens and János Kollár, chaired the organizing committee. Activities were centered around themes, one per month, under the guidance of experts in each area. There were also four short workshops, which attracted many participants and helped considerably in communicating the new directions in algebraic geometry to a large audience.

This book gives a good idea of the intellectual content of the special year and of the workshops. Its articles represent very well the change of direction and branching out witnessed by algebraic geometry in the last few years.

Mathematical Sciences Research Institute
Publications

28

CURRENT TOPICS IN COMPLEX ALGEBRAIC GEOMETRY

Mathematical Sciences Research Institute Publications

Volume 1 Freed and Uhlenbeck: *Instantons and Four–Manifolds*, second edition

Volume 2 Chern (editor): *Seminar on Nonlinear Partial Differential Equations*

Volume 3 Lepowsky, Mandelstam, and Singer (editors): *Vertex Operators in Mathematics and Physics*

Volume 4 Kac (editor): *Infinite Dimensional Groups with Applications*

Volume 5 Blackadar: *K–Theory for Operator Algebras*

Volume 6 Moore (editor): *Group Representations, Ergodic Theory, Operator Algebras, and Mathematical Physics*

Volume 7 Chorin and Majda (editors): *Wave Motion: Theory, Modelling, and Computation*

Volume 8 Gersten (editor): *Essays in Group Theory*

Volume 9 Moore and Schochet: *Global Analysis on Foliated Spaces*

Volume 10 Drasin, Earle, Gehring, Kra, and Marden (editors): *Holomorphic Functions and Moduli I*

Volume 11 Drasin, Earle, Gehring, Kra, and Marden (editors): *Holomorphic Functions and Moduli II*

Volume 12 Ni, Peletier, and Serrin (editors): *Nonlinear Diffusion Equations and Their Equilibrium States I*

Volume 13 Ni, Peletier, and Serrin (editors): *Nonlinear Diffusion Equations and Their Equilibrium States II*

Volume 14 Goodman, de la Harpe, and Jones: *Coxeter Graphs and Towers of Algebras*

Volume 15 Hochster, Huneke, and Sally (editors): *Commutative Algebra*

Volume 16 Ihara, Ribet, and Serre (editors): *Galois Groups over Q*

Volume 17 Concus, Finn, and Hoffman (editors): *Geometric Analysis and Computer Graphics*

Volume 18 Bryant, Chern, Gardner, Goldschmidt, and Griffiths: *Exterior Differential Systems*

Volume 19 Alperin (editor): *Arboreal Group Theory*

Volume 20 Dazord and Weinstein (editors): *Symplectic Geometry, Groupoids, and Integrable Systems*

Volume 21 Moschovakis (editor): *Logic from Computer Science*

Volume 22 Ratiu (editor): *The Geometry of Hamiltonian Systems*

Volume 23 Baumslag and Miller (editors): *Algorithms and Classification in Combinatorial Group Theory*

Volume 24 Montgomery and Small (editors): *Noncommutative Rings*

Volume 25 Akbulut and King: *Topology of Real Algebraic Sets*

Volume 26 Judah, Just, and Woodin (editors): *Set Theory of the Continuum*

Volume 27 Carlsson, Cohen, Hsiang, and Jones (editors): *Algebraic Topology and Its Applications*

Volume 28 Clemens and Kollár (editors): *Current Topics in Complex Algebraic Geometry*

Volumes 1 through 27 are available from Springer-Verlag

Current Topics in Complex Algebraic Geometry

Edited by

Herbert Clemens and János Kollár
University of Utah

Herbert Clemens
János Kollár
Department of Mathematics
University of Utah
Salt Lake City, UT 84112-1107

Mathematical Sciences Research Institute
1000 Centennial Drive
Berkeley, CA 94720

The Mathematical Sciences Research Institute wishes to acknowledge support by the National Science Foundation.

CAMBRIDGE UNIVERSITY PRESS
Cambridge, New York, Melbourne, Madrid, Cape Town,
Singapore, São Paulo, Delhi, Tokyo, Mexico City

Cambridge University Press
The Edinburgh Building, Cambridge CB2 8RU, UK

Published in the United States of America by Cambridge University Press, New York

www.cambridge.org
Information on this title: www.cambridge.org/9781107403840

First published 1995
First paperback edition 2011

A catalogue record for this publication is available from the British Library

[Insert Library of Congress data if available from input material]

ISBN 978-0-521-56244-7 Hardback
ISBN 978-1-107-40384-0 Paperback

Contents

Complex Algebraic Geometry
MSRI Publications
Volume **28**, 1995

Preface

The 1992/93 academic year at the Mathematical Sciences Research Institute was devoted to Complex Algebraic Geometry. The cochairs of the organizing committee were Herb Clemens and myself. Early on we decided that the activities would be centered around themes, each month having its own theme. In organizing each theme, we relied on the help of experts in that area. The success of the year depended very much on their energy and enthusiasm. The following is the list of topics and of the organizers:

September: *Algebraic cycles* (A. Beilinson, W. Fulton)
October: *Vector bundles* (R. Lazarsfeld)
November: *Higher-dimensional geometry* (J. Kollár, S. Mori)
December: *Curves, abelian varieties and their moduli* (A. Beauville, J. Harris)
January: *Surface theory, classical projective geometry* (R. Friedman, J. Harris)
February: *Topology of moduli spaces* (E. Arbarello)
March: *Enumerative and computational algebraic geometry* (W. Fulton)
April: *Crystalline methods and Hodge theory* (A. Beilinson, A. Ogus)
May: *Singularity theory and Hodge theory* (M. Green, J. Steenbrink)

There were also four short workshops, which attracted many participants and helped considerably in communicating the new directions in algebraic geometry to a large audience. The following is the list of workshops held at MSRI:

Algebraic cycles (A. Beilinson, W. Fulton)
Higher dimensional geometry (J. Kollár, S. Mori)
Curves, abelian varieties and their moduli (A. Beauville, J. Harris)
Enumerative geometry and physics (W. Fulton)

In conjunction with the Complex Algebraic Geometry Year at MSRI there were three additional workshops organized at nearby Universities:

Vector bundles, UC Los Angeles (R. Lazarsfeld)
Crystalline methods and Hodge theory, UC Berkeley (A. Ogus)
Hodge theory and singularities, UC Riverside (M. Green, Z. Ran, J. Steenbrink)

We would like to thank the organizers of the workshops for their valuable contributions to the scientific success of the special year.

The aim of this volume is to collect survey articles that give a good idea of the direction of the special year and of the workshops. We were very fortunate to have had the special year at a time when algebraic geometry was undergoing a major change. To put it succinctly, algebraic geometry has opened up to ideas and connections from other fields that have traditionally been far away. The articles of this volume represent this change of direction very well.

(As the reader may have noticed, the initials of all the contributors of this volume fall in the first half of the alphabet. What is the explanation of this overrepresentation? Is this one of the new directions? What, if any, remedies should be applied?)

Arapura surveys recent results connected with the fundamental groups of smooth projective varieties. Traditionally the fundamental group was viewed as a strange invariant of a variety, not related to the usual concepts of algebraic geometry, except in the case of curves. Recently there has been a complete turn-around, and now we understand that the fundamental group relates to diverse questions in subtle ways, many of them still poorly understood. This direction generated a lot of interest at the time of the special year and since. One of the most exciting developments is the connection with the theory of harmonic maps.

Vector bundles on curves are very familiar objects. A few years ago interest in them was rekindled by the Verlinde conjectures coming from theoretical physics. These conjectures give an explicit formula for the Riemann–Roch theorem for the moduli of vector bundles. The intrinsic beauty and accessibility of these formulas suddenly generated a lot of activity. At times it seemed that everyone was trying to understand their behaviour. This high level of interest yielded several solutions from different points of view. The varied approaches turned out to complement each other very nicely. This led to further development in the theory of moduli spaces, where methods from algebraic geometry and analysis of geometric PDE's can be used side by side. This is the subject of the survey article by Beauville.

One of the major internal developments of algebraic geometry in the last decade has been the emergence of higher-dimensional birational geometry, and especially Mori's minimal model program. After a very intense decade of perfecting various techniques, the direction of the field turned toward applying the ideas and methods of the program to other questions. The article by Corti reviews the main steps and techniques of the program, and then goes on to explain recent applications to the study of the moduli of surfaces, to the Nöther–Fano–Iskovskikh theory of birational maps, and to the effective base point freeness question.

The latter topic is taken up in more detail in the survey by Ein. This is an especially well chosen topic, since here one can illustrate the general methods of the minimal model program in a very simple and clear situation. These applications are rather surprising and they opened up many new directions. Recent extensions of the theory to vector bundles hold a lot of promise.

Abelian varieties are one of the oldest and most enduring topics of algebraic geometry, and this field always comes up with new perspectives. The short survey by Debarre guides the reader to many of them. One of the most surprising new developments is the realization that the lattices of Jacobians have very strong metric properties. It still remains to be seen how these can be understood in terms of algebraic geometry.

Spectral covers originated in the study of vector bundles of curves. They have found connections with the study of vector bundles on higher-dimensional varieties, with the theory of fundamental groups of varieties and with integrable Hamiltonian systems. The survey by Donagi is devoted to their study.

Another classical topic that always finds a new life is the theory of the moduli of curves. The study of this moduli space can be approached from many angles. The connection with the Torelli group is the guiding principle of the article by Hain. This method reveals profound connections with the topology of Riemann surfaces, with the cohomology theory of discrete groups and with Hodge theory.

The last contribution of the volume by Mukai provides a connection with many topics. The theory of special vector bundles on curves can be viewed as a natural development starting with questions about the moduli of curves. It also relates to the questions of Verlinde type. Most surprisingly, several of the resulting moduli spaces give interesting examples of Fano varieties. This led to a deeper understanding of the classification of Fano threefolds.

We hope that, by collecting these lectures in a single volume, we can convey at least some of the sense of the new directions that started to develop during the special year at MSRI.

János Kollár
Department of Mathematics
University of Utah
Salt Lake City, UT 84112
E-mail address: kollar@math.utah.edu

Complex Algebraic Geometry
MSRI Publications
Volume **28**, 1995

Fundamental Groups of Smooth Projective Varieties

DONU ARAPURA

To the memory of Boris Moishezon

ABSTRACT. This article is a brief survey of work related to the structure of topological fundamental groups of complex smooth projective varieties.

These notes, which are based on a talk given at MSRI in April 1993, are intended as a brief guide to some recent work on fundamental groups of varieties. For the most part, I have just tried to explain the results (often in nonoptimal form) and give a few simple examples to illustrate their use. Proofs are either sketched or omitted entirely. The basic question that will concern us is:

Which groups can be fundamental groups of smooth projective varieties?

This is certainly of importance in the topological study of algebraic varieties, but it is also linked to broader issues in algebraic geometry. Let us call the class of such groups $\mathcal{P}$. As an application of the Lefschetz hyperplane theorem [Mi], we can see that any group in $\mathcal{P}$ is the fundamental group of an algebraic surface, and in fact we can even arrange the surface to have general type. Thus failure to answer this question can be viewed as an obstruction to completely classifying algebraic surfaces (even up to homotopy).

I should mention that there is also a nice survey article by Johnson and Rees [JR2] that reviews much of the work done on this problem prior to 1990. My own view of the subject has been shaped, to a large extent, by conversations and correspondence with many people, of whom I would especially like to mention Paul Bressler, Jim Carlson, Dick Hain, János Kollár, Madhav Nori, Mohan Ramachandran and Domingo Toledo. My thanks to F. Campana for catching a silly mistake in an earlier version of these notes.

The terminology used here is fairly standard. The only thing that could cause confusion is that I will say that a group G is an extension of B by A if it fits into an exact sequence:

$$1 \to A \to G \to B \to 1,$$

and not the other way around.

1. Positive results

Since most of the known results are in the negative direction, let us start with some positive ones.

(A$^+$) (Serre) Any finite group lies in $\mathcal{P}$. What Serre [S, Proposition 15] in fact proves is that any finite group acts without fixed points on some smooth complete intersection of any prescribed dimension. Since, by Lefschetz's hyperplane theorem, complete intersections of dimension at least 2 are simply connected, the first statement follows.

(B$^+$) $\mathcal{P}$ is closed under finite products because the class of projective varieties is.

(C$^+$) If $G \in \mathcal{P}$, any subgroup of finite index lies in $\mathcal{P}$, because a finite-sheeted covering of a smooth projective variety can be given the structure of a smooth projective variety.

(D$^+$) As we shall see later, the converse of (C$^+$) is false; however, a weak form (needed below) does hold. Suppose that X is a simply connected complex manifold on which a group G acts faithfully, biholomorphically and properly discontinuously. Assume furthermore that a finite index subgroup $H \subseteq G$ acts freely on X and that the quotient is a projective variety. Then $G \in \mathcal{P}$.

Proof (Kollár). We can assume that H is normal, since otherwise we can replace it by a stabilizer of a coset in G/H. Let S be a smooth projective variety with fundamental group G/H, and let $\tilde{S}$ be its universal cover. Then the diagonal action of G on $X \times \tilde{S}$ is free, so the quotient is smooth and has G as its fundamental group. Furthermore $(X \times \tilde{S})/G$ is a projective variety since it possesses a finite holomorphic map to $X/G \times S$. □

(E$^+$) For any positive integer g, the group

$$\langle a_1, a_2, \ldots, a_{2g} \mid [a_1, a_{g+1}], \ldots, [a_g, a_{2g}] = 1 \rangle,$$

which is the fundamental group of a curve of genus g, lies in $\mathcal{P}$.

(F$^+$) If G is a semisimple Lie group such that the quotient D of G by a maximal compact subgroup is a Hermitian symmetric space of noncompact type, any cocompact discrete subgroup $\Gamma \subset G$ lies in $\mathcal{P}$. When Γ is torsion-free, this follows from the fact that G acts freely on D and the quotient when endowed with the Bergman metric satisfies the conditions of Kodaira's embedding theorem (see [H, ch. VIII] and [KM, p. 144]), and is consequently a smooth projective variety with fundamental group Γ. In the general case, Γ contains a torsion-free subgroup of finite index [Sel, Lemma 9], so we can appeal to (D^+). Two simple examples to keep in mind are: $G = \mathrm{SU}(n,1)$ the group of unimodular matrices preserving the indefinite form $|z_1|^2 + \cdots + |z_n|^2 - |z_{n+1}|^2$, and $G = \mathrm{Sp}(2n, \mathbb{R})$ the group of matrices preserving the standard symplectic form on $\mathbb{R}^{2n}$; maximal compact subgroups are given by

$$K_1 = \left\{ \begin{pmatrix} A & 0 \\ 0 & \det A^{-1} \end{pmatrix} \mid A \in U(n) \right\}$$

and

$$K_2 = \left\{ \begin{pmatrix} X & Y \\ -Y & X \end{pmatrix} \mid X + iY \in U(n) \right\},$$

respectively.

$\mathrm{SU}(n,1)/K_1$ can be identified with the unit ball B^n in $\mathbb{C}^n$, and $\mathrm{Sp}(2n,\mathbb{R})/K_2$ can be identified with the Siegel upper half plane $\mathcal{H}_n$ of $n \times n$ symmetric matrices with positive definite imaginary part. In particular, when $n = 1$, we see that all the groups described in (D^+) with $g \geq 2$ arise in this fashion. Although it is difficult to construct cocompact lattices explicitly, they always exist for any G [Bo].

($\mathbf{G}^+$) (Toledo [T1]) In the above situation, suppose that G is the group of real points of an algebraic group defined over $\mathbb{Q}$. Assume that none of the irreducible factors of D is isomorphic to B^1, B^2, or $\mathcal{H}_2$. Then any arithmetic subgroup of G lies in $\mathcal{P}$. In particular, $\mathrm{Sp}(2n, \mathbb{Z}) \in \mathcal{P}$ when $n > 2$. I will describe the idea in this example. As in the proof of (D^+), we can obtain a free action of $\mathrm{Sp}(2n, \mathbb{Z})$ on the product of $X = \mathcal{H}_n$ and a suitably chosen simply connected smooth projective variety $\tilde{S}$. The variety

$$Y = (X \times \tilde{S}) / \mathrm{Sp}(2n, \mathbb{Z})$$

is not projective but only quasiprojective. It has a compactification $\bar{Y}$ obtained by normalizing $\bar{A}_n \times S$ in the function field of Y, where $\bar{A}_n$ is the Satake compactification of $A_n = X/\mathrm{Sp}(2n,\mathbb{Z})$. The variety $\bar{Y}$ is projective and we will fix a projective embedding. The codimension of the complement $\bar{Y} - Y$ is at least 3; therefore we can slice $\bar{Y}$ by hyperplanes, in general position, until we get a smooth projective surface Z contained in Y. A strong form of the Lefschetz theorem, due to Goresky and Macpherson, guarantees that the fundamental group of Z is isomorphic to that of Y, which is of course $\mathrm{Sp}(2n, \mathbb{Z})$.

($\mathbf{H}^+$) Toledo [T2] has solved the long-outstanding problem of showing that $\mathcal{P}$ contains nonresidually finite groups (i.e., groups that don't embed into their profinite completions). Other examples have since been constructed by Catanese,

Kollár and Nori (see [CK]). The simplest such example is the preimage of $\mathrm{Sp}(6,\mathbb{Z})$ in the connected 3-fold cyclic cover of $\mathrm{Sp}(6,\mathbb{R})$. The nonresidual finiteness of this and related groups is due to Deligne [De].

($\mathbf{I}^+$) Sommese and Van de Ven [SV], and later Campana [Ca], have shown that $\mathcal{P}$ contains nonabelian torsion-free nilpotent groups, and this contradicts a long-held belief by many workers in the area including this author. (Unfortunately the belief and the counterexample in [SV] existed concurrently for quite some time.) The examples are constructed as follows: Choose an abelian n-fold A and a finite map to $\mathbb{P}P^n$. Let X be the preimage in A of a generic translate of an abelian d-fold in $\mathbb{P}P^n$ with $d \geq 2$. Then a suitable double cover of X has as fundamental group a nonsplit central extension of an abelian group by $\mathbb{Z}$. Sommese and Van de Ven used a specific choice $(n,d) = (4,2)$, and this yields an extension of $\mathbb{Z}^{12}$ by $\mathbb{Z}$ in $\mathcal{P}$.

2. Simple obstructions

Now let us consider the various known obstructions for a group to lie in $\mathcal{P}$. The first obvious constraint, coming from the fact that any variety admits a finite triangulation, is that the groups in $\mathcal{P}$ are finitely presented. A more subtle constraint comes from Hodge theory, which implies that the first Betti number of a smooth projective variety is even (see for example [GH, p. 117] or [KM, p. 115]). Therefore, by Hurewicz's theorem:

($\mathbf{A}^-$) If $\mathrm{rank}(G/[G,G])$ is odd, then $G \notin \mathcal{P}$.

By virtue of (C^+) we obtain a strengthening:

($\mathbf{A}'^-$) If G has a subgroup H of finite index with $\mathrm{rank}(H/[H,H])$ odd, then $G \notin \mathcal{P}$.

In particular, *the free group on n generators F_n is not in $\mathcal{P}$*. This is immediate from (A^-) if n is odd. If n is even, let $g_1, g_2, \ldots, g_n$ be free generators of F_n. Then the subgroup generated by

$$g_1^2,\ g_2,\ g_3,\ \ldots,\ g_n,\ g_1g_2g_1^{-1},\ g_1g_3g_1^{-1},\ \ldots,\ g_1g_ng_1^{-1}$$

is a subgroup of finite index that is free on an odd number of generators. This can be seen geometrically as follows. Let X be a bouquet of n circles; its fundamental group is F_n, one generator g_i for each circle. Let $\tilde{X} \to X$ be the $\mathbb{Z}/2\mathbb{Z}$-covering corresponding to the homomorphism $F_n \to \mathbb{Z}/2\mathbb{Z}$ defined by sending g_1 to 1 and the other generators to 0. Then the image of $\pi_1(\tilde{X})$ in $\pi_1(X)$ is the subgroup defined above and the essential loops in $\tilde{X}$ correspond to the given generators.

As a second example, let G be the semidirect product of $\mathbb{Z}^2$ with $\mathbb{Z}/2\mathbb{Z}$, where the second group acts on the first through the involution $(x,y) \mapsto (x,-y)$. Then $G \notin \mathcal{P}$ since $\mathrm{rank}(G/[G,G]) = 1$. This example shows that the converse of (C^+)

is false. It may also be worth remarking that this example is the fundamental group of the Klein bottle.

At this point one might be tempted to speculate that (A$'^-$) is the only obstruction. But now let's consider a subtler example, the 3×3 Heisenberg group H, which is the group of 3×3 upper triangular integer matrices with 1's on the diagonal. The rank of the abelianization of any finite index subgroup is 2. Nevertheless, $H \notin \mathcal{P}$. We will give several proofs of this. The first, due to Johnson and Rees [JR1], is the simplest. Before indicating the argument, we will recall a few facts about group cohomology (details can be found in [Br]). If M is an abelian group upon which a group G acts, then the cohomology group $H^i(G, M)$ can be defined in an entirely algebraic manner, either explicitly in terms of cocycles or more abstractly via derived functors. When $M = \mathbb{R}$ with trivial G action, $H^*(G, \mathbb{R})$ becomes a graded ring under cup product. If X is a connected topological space, there is a natural ring homomorphism

$$H^*(\pi_1(X), \mathbb{R}) \to H^*(X, \mathbb{R}),$$

which is an isomorphism when $* = 1$, and is an isomorphism for all $*$ provided that the universal cover of X is contractible, in which case X is called a $K(\pi_1(X), 1)$.

(B$^-$) (Johnson–Rees) If $H^1(G, \mathbb{R}) \neq 0$ and the map $\mathrm{sq}_G : \wedge^2 H^1(G, \mathbb{R}) \to H^2(G, \mathbb{R})$ induced by cup product vanishes, then $G \notin \mathcal{P}$.

PROOF. Suppose $G = \pi_1(X)$, where X is a smooth projective n-dimensional variety. Then there is an isomorphism $H^1(G, \mathbb{R}) \cong H^1(X, \mathbb{R})$ and sq_G factors through $\wedge^2 H^1(X, \mathbb{R}) \to H^2(X, \mathbb{R})$, so it is enough to check that this is nonzero. Given a nonzero class $\alpha \in H^1(X)$, there exists by Poincaré duality a $\beta \in H^{2n-1}(X)$ such that $\alpha \cup \beta \neq 0$. By the Hard Lefschetz theorem [GH, p. 122] we have $\beta = \gamma \cup L^{n-1}$, where $\gamma \in H^1(X)$ and L is the class of a hyperplane section. Therefore $\alpha \cup \gamma \neq 0$. □

The cohomology ring of H can be easily computed topologically once one observes that a $K(H, 1)$ is given by

$$\left\{ \begin{pmatrix} 1 & x & z \\ 0 & 1 & y \\ 0 & 0 & 1 \end{pmatrix} \mid x, y, z \in R \right\} / H.$$

The first and second cohomology groups H are isomorphic to $\mathbb{R}[dx] \oplus \mathbb{R}[dy]$ and $\mathbb{R}[dx \wedge dz] \oplus \mathbb{R}[dy \wedge dz]$; in particular $\mathrm{sq}_H = 0$.

3. Groups with more than one end

The fact that F_n is not in $\mathcal{P}$ is part of a more general phenomenon that we will explain in this section. If X is a topological space, for any subset K let

$E(K)$ be the set of connected components of $X - K$ with noncompact closure. The number of ends of X is defined as

$$\sup\{\#E(K) \mid K \subseteq X \text{ compact}\} \in \mathbb{N} \cup \{\infty\}.$$

Exercise: show that $\mathbb{C}$ has one end and $\mathbb{R}$ has two.

The number of ends of a finitely generated group G can be defined in a purely group-theoretic fashion as the dimension of a cohomology group

$$1 + \dim H^1(G, \mathbb{Z}/2\mathbb{Z}[G]).$$

However, it has a more geometric interpretation. Suppose that X is a simplicial complex upon which G acts freely and simplicially with compact (i.e., finite) quotient. Then the number of ends of G and X coincide. So, for example, as a corollary of the exercise: $\mathbb{Z}^2$ has one end and $\mathbb{Z}$ has two.

Given a finite set of generators, there is a classical method for building a space upon which G acts, namely the Cayley graph: the vertices are the elements of the group and two vertices are connected by an edge if one vertex can be obtained from the other by multiplication (on the right, say) by a generator or an inverse of one. This has an obvious left action by G with compact quotient; thus the number of ends of G and its graph are the same. Let's consider F_n with its standard generators. Its Cayley graph is an infinite tree where $2n$ branches emanate from any vertex. Clearly this space has infinitely many ends when $n > 1$. More generally, any free product with nontrivial factors other than $\mathbb{Z}/2\mathbb{Z} * \mathbb{Z}/2\mathbb{Z}$ has infinitely many ends; the exceptional case has two. The converse of the previous sentence is very close to being true, thanks to a deep theorem of Stallings; see [SW, Section 6] for the precise statement.

Building on work of Gromov [G], Bressler, Ramachandran and the author [ABR] have obtained:

(C$^-$) If a group has more than one end, it does not lie in $\mathcal{P}$. An extension of a group with infinitely many ends by a finitely generated group does not lie in $\mathcal{P}$.

The first statement implies Gromov's result [G] (see also [JR1]) that $\mathcal{P}$ contains no nontrivial free products. The Heisenberg group H has one end and does not surject onto a group with infinitely many ends, so it is not covered by the above result. As an application of the last part of the above criterion, let's show that the *braid group* B_n is not in $\mathcal{P}$. This group can be defined as the fundamental group of the space of n distinct unordered points in the plane $\mathbb{R}^2$. For other definitions and basic properties see [Bi]. Since $B_2 = \mathbb{Z}$, it cannot lie in $\mathcal{P}$. The standard presentation for B_3 is

$$\langle s_1, s_2 \mid s_1 s_2 s_1 = s_2 s_1 s_2 \rangle.$$

If we set $a = s_1s_2$ and $b = s_1s_2s_1$, we get a new presentation $B_3 = \langle a, b \mid a^3 = b^2\rangle$. The subgroup $N = \langle a^3\rangle$ is normal and

$$B_3/N \cong \mathbb{Z}/2\mathbb{Z} * \mathbb{Z}/3\mathbb{Z}$$

has infinitely many ends; consequently $B_3 \notin \mathcal{P}$. Let P_n be the pure braid group. This is the fundamental group of the space X_n of n ordered points in $\mathbb{R}^2$. It is a subgroup of finite index in B_n, so it suffices to show that $P_n \notin \mathcal{P}$ when $n > 3$. Note that the image of P_3 in B_3/N has infinitely many ends because this property is stable under passage to subgroups of finite index. The projection $X_n \to X_3$ is a fiber bundle and its fiber is homotopic to a finite complex. Thus there is a surjection $p : P_n \to P_3$ with finitely generated kernel. Therefore $p^{-1}(N \cap P_3) \subset P_n$ is a finitely generated normal subgroup such that the quotient has infinitely many ends.

4. Rational homotopy

One of the key insights coming from rational homotopy theory is that the algebra of differential forms on a manifold contains a lot more topological information than just the cohomology ring. One can obtain information about all nilpotent quotients of the fundamental group (not just abelian ones). This information can be systematized by replacing G by the inverse limit of all of its nilpotent quotients:

$$\hat{G} = \lim_{\leftarrow} N,$$

which is called its *nilpotent completion.* We would like to introduce a coarser construction, the Malcev (or rational nilpotent) completion $\mathcal{G}$, which should be thought of as "$\hat{G} \otimes \mathbb{Q}$". Malcev has shown that any finitely generated torsion-free nilpotent group can be embedded as a Zariski dense subgroup of a unipotent linear algebraic group (i.e., an algebraic subgroup of a group of upper triangular matrices) over $\mathbb{Q}$. This leads us to define

$$\mathcal{G} = \lim_{\leftarrow} U,$$

where the limit runs over all representations of G into unipotent algebraic groups U defined over $\mathbb{Q}$. As a trivial but instructive example, let G be abelian. Then $\hat{G} = G$ and $\mathcal{G}$ really is $G \otimes \mathbb{Q}$. The prounipotent group $\mathcal{G}$ is completely determined by its Lie algebra:

$$L(G) = \mathrm{Lie}(\mathcal{G}) = \lim_{\leftarrow} \mathrm{Lie}(U),$$

and this is usually more convenient to work with. $L(G)$ can be topologized by taking the above inverse limit in the category of topological Lie algebras, where each factor $\mathrm{Lie}(U)$ is equipped with the discrete topology. These definitions, while efficient, have the disadvantage of making $\mathcal{G}$ and $L(G)$ seem mysterious;

they aren't. They can be realized quite explicitly as subsets of the completion of the group ring $\mathbb{Q}[G]$ at its augmentation ideal [Q, Appendix A].

Let's work out two examples. First we will compute $L(H)$ for the 3×3 Heisenberg group. We take U to be the unipotent group of upper triangular 3×3 rational matrices. Then the map $H \to U$ is an initial object in the above inverse system; therefore $\mathcal{G} = U$ and $L(H) = \mathrm{Lie}(U)$, the Lie algebra of strictly upper triangular 3×3 rational matrices. Next consider the free group F_n on n generators $X_1, \ldots, X_n$. Let FL_n be the free $\mathbb{Q}$-Lie algebra on n generators $x_1, \ldots, x_n$. Let

$$C^N \mathrm{FL}_n = [\mathrm{FL}_n, [\mathrm{FL}_n, \ldots, [\mathrm{FL}_n, \mathrm{FL}_n], \ldots]] \qquad (N+1 \ \mathrm{FL}_n\text{'s})$$

be the N-th term of the lower central series. Then it can be checked that $L(F_n)$ is the completion

$$\widehat{\mathrm{FL}}_n = \lim_{\overleftarrow{N}} (\mathrm{FL}_n / C^N \mathrm{FL}_n)$$

of the free Lie algebra with respect to the topology determined by the lower central series. The main point is that

$$X_i \mapsto \exp(\mathrm{ad}(x_i)) \in \mathrm{GL}(\mathrm{FL}_n / C^N \mathrm{FL}_n)$$

determines a cofinal family of unipotent representations of F_n.

We would like to describe the Lie algebra $L(G)$, for arbitrary G, in terms of generators and relations. For generators, choose $x_1, x_2, \ldots, x_n \in L(G)$ such that they determine a basis of

$$L(G)/[L(G), L(G)] \cong (G/[G,G]) \otimes \mathbb{Q}.$$

These elements will generate a dense subalgebra of $L(G)$; thus we obtain a continuous surjective homomorphism $\widehat{\mathrm{FL}}_n \to L(G)$. Call the kernel $I(G)$ (we'll suppress the dependence on the x_i). Any element of $\widehat{\mathrm{FL}}_n$ can be expanded as an infinite series

$$\sum a_i x_i + \sum b_{ij}[x_i, x_j] + \sum c_{ijk}[x_i, [x_j, x_k]] + \cdots .$$

The degree of the element is the degree of the smallest term—in other words, the length of the shortest commutator appearing in the series. Let $I_2(G)$ be the closed ideal of $\widehat{\mathrm{FL}}_n$ generated by elements of $I(G)$ of degree 2. The following result of Deligne, Griffiths, Morgan, and Sullivan is the first deep result in this area. (The result as stated here does not appear explicitly in their paper [DGMS], but it is a well known consequence of it. See [CT2] and [M, Section 9, 10] for further discussion.)

($\mathbf{D}^-$) If for some (any) choice of generators, $I(G) \neq I_2(G)$ then $G \notin \mathcal{P}$.

Or, in plain English, the essential relations of $L(G)$, for $G \in \mathcal{P}$, are quadratic. One can even say what they are: they are dual (in a natural sense) to the kernel of the map sq_G of Section 2. For example, let G be the fundamental group of a smooth projective curve of genus g. Then $L(G)$ is isomorphic to the quotient of $\widehat{\mathrm{FL}}_{2g}$ by the quadratic relation

$$\sum_{i=1}^{g} [x_i, x_{i+g}] = 0.$$

Consider the 3×3 Heisenberg group. We can choose two generators x_1, x_2 of $L(H)$ corresponding to the matrices with 1's at $(1,2)$ and $(2,3)$, respectively, and zeros elsewhere. Then clearly there aren't any quadratic relations, although there are cubic ones. Thus again we conclude that $H \notin \mathcal{P}$. This sort of reasoning allows one to eliminate a lot of nilpotent groups from $\mathcal{P}$, although even for this class the method does not yield a definitive answer. Carlson and Toledo have pointed out to me that the Lie algebra of the 5×5 Heisenberg group does in fact have quadratic relations; nevertheless it doesn't lie in $\mathcal{P}$ for other reasons [CT2]. In fact, there are many examples of groups $G \notin \mathcal{P}$ for which $L(G)$ has quadratic relations—for instance, F_n, and less trivially P_n (see [Ko]).

5. Representation varieties

Let Γ be a group generated by finitely many elements $\gamma_1, \ldots, \gamma_n$. Giving a representation of Γ into G is the same thing as choosing n elements $g_i \in G$ satisfying the relations satisfied by γ_i. Thus, if G is a real algebraic group, the set of representations $\mathrm{Hom}(\Gamma, G)$ carries the structure of a (possibly reducible) real algebraic variety. When $\Gamma \in \mathcal{P}$ the local structure of this variety is well understood, thanks to the work of Deligne, Goldman, Milson [GM] and Simpson [S1]:

($\mathbf{E}^-$) If $\Gamma \in \mathcal{P}$, then $X = \mathrm{Hom}(\Gamma, G)$ has quadratic singularities at points corresponding to semisimple representations. More precisely, the completion of the local ring of X at such a point is isomorphic to the quotient of a formal power series ring by an ideal generated by quadratic polynomials.

There is a strong formal similarity between these results and (D^-), and in fact the proof uses a generalization of the methods of (D^-) to local systems. For the third time, let's look at the Heisenberg group H. Let G be the group of 3×3 upper triangular real matrices. Then the singularity of $\mathrm{Hom}(H, G)$ at the trivial representation can be shown to be a cubic cone. Thus $H \notin \mathcal{P}$.

6. Lattices in Lie groups

While it seems very difficult to characterize all groups in $\mathcal{P}$, a more reasonable problem would be to classify the discrete subgroups of Lie groups that lie in $\mathcal{P}$. In the first section we indicated some positive results in this direction; now we consider some obstructions. But first some terminology. A lattice of a Lie group is a discrete subgroup such that the quotient has finite volume with respect to Haar measure (this is certainly the case when the quotient is compact, for example). $\mathrm{SO}(n,1)$ is the group of unimodular matrices preserving the form $x_1^2 + \cdots + x_n^2 - x_{n+1}^2$. Using techniques from the theory of harmonic maps, Carlson and Toledo [CT] obtain:

(F$^-$) No cocompact discrete subgroup of $\mathrm{SO}(n,1)$, with $n > 2$, lies in $\mathcal{P}$.

Note that the symmetric space associated to $\mathrm{SO}(n,1)$ is not Hermitian when $n > 2$. In fact these authors conjecture that a cocompact lattice in a semisimple group is never in $\mathcal{P}$ unless the associated symmetric space is Hermitian. As further evidence, Carlson and Hernandez [CH] show that a lattice in the automorphism group of the Cayley plane does not lie in $\mathcal{P}$.

The strongest results of this sort have been obtained by Simpson [S1] (see also [C]). To state them, we need some more terminology. Let W be a real algebraic group, G the associated complex group and σ the complex conjugation of G corresponding to W. A Cartan involution is an automorphism C of G such that $C^2 = 1$, $\tau = C\sigma = \sigma C$, and the set of τ-fixed points of G is a compact group that meets every component G. The group W is of Hodge type if there is a γ in the identity component of G such that $x \mapsto \gamma x \gamma^{-1}$ is a Cartan involution (this is equivalent to Simpson's definition [S1, Section 4.42]). For example, $\mathrm{Sp}(2n,\mathbb{R})$ is of Hodge type, for we can take

$$\gamma = \begin{pmatrix} 0 & I \\ -I & 0 \end{pmatrix}.$$

The group of τ-fixed points on the associated complex group $\mathrm{Sp}(2n,\mathbb{C})$ is

$$\mathrm{Sp}(2n,\mathbb{C}) \cap \mathrm{SU}(2n).$$

A list of simple groups of Hodge type can be found in [S1, pp. 50-51]. $\mathrm{SL}_n(\mathbb{R})$ is not of Hodge type as soon as $n \geq 3$. The most important (and motivating) examples of groups of Hodge type are the Zariski closures of the monodromy groups of complex variations of Hodge structure. This gives another explanation of why $\mathrm{Sp}(2n,\mathbb{R})$ is of Hodge type: namely, it is the (real) Zariski closure of the monodromy group of the variation of Hodge structure associated to the first cohomology of a family of n-dimensional abelian varieties. Simpson shows that any representation of the fundamental group of a smooth projective variety into a reductive group can be deformed into one coming from a variation of

Hodge structure. In particular, if the given representation is rigid, in the sense that every nearby representation is conjugate to it, then it must be conjugate to one coming from a variation of Hodge structure. Thus if the image of the representation is Zariski dense (which would be the case if it were a lattice) the target group must be of Hodge type. Therefore:

(G⁻) If Γ is a rigid lattice in a reductive real algebraic group that is not of Hodge type, $\Gamma \notin \mathcal{P}$.

To apply this, one needs to be able to check that a given lattice is rigid. Fortunately there are a number of rigidity theorems in existence; they are summarized in [S1, p. 53]. In particular $\mathrm{SL}_n(\mathbb{Z})$ is rigid when $n \geq 3$, so it cannot lie in $\mathcal{P}$. (Neither does $\mathrm{SL}_2(\mathbb{Z})$, but one must appeal to (C⁻) in this case.) Note also that $\mathrm{SO}(n,1)$ is of Hodge type when n is even, so that (G⁻) does not imply (F⁻).

The above results were concerned with lattices in reductive groups. Now let's look at opposite end of the spectrum, namely lattices in nilpotent or, more generally, solvable Lie groups. A lattice in a nilpotent Lie group is finitely generated and nilpotent as a discrete group, and conversely any finitely generated torsion-free nilpotent group is such a lattice by Malcev's theorem. We have already mentioned in (I⁺) that it had been expected that such groups cannot lie in $\mathcal{P}$ unless they are abelian. At the moment the situation regarding which nilpotent groups lie in $\mathcal{P}$ is somewhat murky, but hopefully a clearer picture will emerge with time. For the state of the art, see [CT2] and [CT3].

Now let us turn to solvable groups. A lattice in a solvable Lie group is solvable as a discrete group, and moreover satisfies a noetherian condition that all of its subgroups are finitely generated. A group satisfying these conditions is called polycyclic (we're taking this as the definition for simplicity; the usual one is different although equivalent). Nori and the author [AN] have shown that the problem of understanding polycyclic groups in $\mathcal{P}$ reduces to the previous one:

(H⁻) If a polycyclic group lies in $\mathcal{P}$, it contains a nilpotent group of finite index.

Let's consider the following example. An action of $\mathbb{Z}^2$ on itself is determined by specifying two commuting automorphisms $T_i \in \mathrm{GL}_2(\mathbb{Z})$. For simplicity, let $T_2 = I$ and let G be the semidirect product of $\mathbb{Z}^2$ with itself using this action. Then G is polycyclic. Using (H⁻) and (A⁻), one can check that $G \notin \mathcal{P}$ unless T_1 has finite order. When T_1 has finite order, it is not too hard to see that G can be realized as the fundamental group of one of the bielliptic surfaces listed in [Be, pp. 113-114]. Thus $G \in \mathcal{P}$.

7. Maps to curves

In recent years a number of remarkable results have been obtained that fit into the following general pattern:

If the fundamental group of smooth projective variety satisfies some suitable hypothesis (say $\diamondsuit$) the variety maps onto curve of genus 2 or more.

In keeping with the aims of these notes, we will concentrate on the group-theoretic consequences. For the sake of expedience let's call a group "curve-dominating" if it possesses a surjective homomorphism onto

$$\Gamma_g = \langle a_1, a_2, \ldots, a_{2g} \mid [a_1, a_{g+1}], \ldots, [a_g, a_{2g}] = 1 \rangle$$

for some $g \geq 2$. Thus any group in $\mathcal{P}$ satisfying ($\diamondsuit$) must be curve-dominating. Now let's look at some specific instances.

In Section 3 we saw that $\mathcal{P}$ does not contain nontrivial free products. However, it does contain amalgamated free products. For example, Γ_{g+h} is a free product of Γ_g and Γ_h amalgamated over $\mathbb{Z}$ (this can seen by decomposing a curve of genus $g+h$ into a connected sum of curves of genus g and h). Call an amalgamated free product $G_1 *_K G_2$ *proper* if the index of K is at least 2 in one of the factors and 3 in the other. Gromov and Schoen [GS, 9.1] have proved:

(I$^-$) Any proper amalgamated free product contained in $\mathcal{P}$ is curve-dominating.

Put another way: a proper amalgamated free product cannot lie in $\mathcal{P}$ unless it's curve-dominating. For example, if G and H are perfect (equal to their commutator subgroups), a proper amalgamated free product $G*_K H$ can never lie in $\mathcal{P}$. Conversely, Mohan Ramachandran has pointed out to me that any curve-dominating group is a proper amalgamated free product. Thus this characterizes curve-dominating elements in $\mathcal{P}$. To see this, let $G \to \Gamma_{g+h}$ be a surjection with $g, h \geq 1$. If we decompose Γ_{g+h} into an amalgamated free product as above, we obtain

$$G \cong G_1 *_K G_2,$$

where G_1, G_2 and K are the preimages of Γ_g, Γ_h and $\mathbb{Z}$ respectively.

The second result of this kind is due, independently, to Green and Lazarsfeld [GL] and Gromov [G]:

(J$^-$) Let $G \in \mathcal{P}$ have a presentation with at least two more generators than relations. Then G is curve-dominating; in fact it surjects onto Γ_g, where g is the rank of the abelianization of G.

This puts a strong restriction on the groups in $\mathcal{P}$ with a small number of relations. We have already seen than that $\mathcal{P}$ contains no groups without relations (free groups), but $\mathcal{P}$ does contain groups with a single defining relation, namely the Γ_g's. Are there any others? I will not venture to guess at this point; however, the previous results yield some nice restrictions.

LEMMA. *Suppose that*

$$G = \langle x_1, x_2, \ldots, x_n \mid R(x_1, \ldots, x_n) \rangle \in \mathcal{P}$$

with $n > 2$. Then

1) *n is even.*
2) *Each x_i occurs at least once in the word R and the number of occurrences of x_i and x_i^{-1} coincide.*
3) *G surjects onto Γ_g with $g = n/2$.*

PROOF. (B^-) implies that $H^2(G, \mathbb{R}) \neq 0$. By a theorem of Lyndon [L, 11.4], this cohomology group is nonzero if and only if each exponent $e_i = 0$, where e_i is the difference between the number of occurrences of x_i and x_i^{-1} in R. Therefore the image of R in the abelianization vanishes. Consequently n is the rank of the abelianization, and so (1) and (3) follow from (A^-) and (J^-). Finally, if some x_i did not occur in R, then G could be decomposed into a free product of the groups generated by x_i and by the remaining variables. But this would contradict the results of Section 3. □

To bring this section to a close, we give a necessary and almost sufficient homological condition for a group in $\mathcal{P}$ to be curve-dominating. For any group G, we recall that

$$H_1(G', \mathbb{Q}) = \frac{G'}{G''} \otimes \mathbb{Q},$$

where $G' = [G, G]$.

($\mathbf{K}^-$) If $G \in \mathcal{P}$ is curve-dominating then

$$\dim H_1(G', \mathbb{Q}) = \infty.$$

Conversely, if this space is infinite-dimensional, G contains a curve-dominating subgroup of finite index.

This follows from the corollary in [A, p. 313]. (Actually, the corollary is misstated there. It should read: if X maps onto a curve of genus at least two then $\dim H_1(\pi_1(X)', \mathbb{Q}) = \infty$, and if $\dim H_1(\pi_1(X)', \mathbb{Q}) = \infty$ then there is a finite sheeted abelian cover that maps onto a curve of genus at least two.)

8. Complex manifolds

In this final section we will indicate the situation for general complex manifolds. For arbitrary manifolds there are no restrictions: Taubes [Ta] has shown that any finitely presented group can occur as the fundamental group of some three-dimensional compact complex manifold. The story for Kähler manifolds is almost identical to the algebraic one; in fact all the results stated previously are known to show that a given group is (or is not) the fundamental group of a compact Kähler manifold. Such groups are called *Kähler groups*; Clearly any group in $\mathcal{P}$ is a Kähler group. The converse is unknown, and would be very hard to disprove. If one were to try, it would be a good idea to first discover a property about $\mathcal{P}$ that is proved by methods peculiar to projective algebraic geometry (such as slicing by hyperplanes, or changing the field of definition). We will mention one such property, whose proof will appear in [AN], which gives a strengthening of (H^-). The proof is arithmetic in nature and we don't know whether the result holds for Kähler groups.

(L^-) If $G \in \mathcal{P}$ is a solvable group that admits a faithful finite dimensional representation over a field of characteristic 0, then G contains a nilpotent subgroup of finite index.

Let's look at a concrete example. Let $p \in \mathbb{Z}$ be a prime and let G be the group of matrices of the form

$$\begin{pmatrix} 1 & * & * & * \\ 0 & * & * & * \\ 0 & 0 & * & * \\ 0 & 0 & 0 & 1 \end{pmatrix},$$

where the entries lie in $\mathbb{Z}[1/p]$ with units on the diagonal. G turns out to be finitely presented [Ab]; however, it contains no nilpotent groups of finite index, so $G \notin \mathcal{P}$. This can also be deduced from a theorem of Simpson [S2] that implies the validity of a conjecture of Beauville and Catanese for smooth projective varieties. A weak form of this conjecture states that the set of line bundles with nonzero first cohomology in Pic^0 of a variety is either infinite or consists of torsion points. It can be checked that, were a smooth projective variety with fundamental group G to exist, it would be a counterexample to this conjecture. Neither argument applies to nonalgebraic manifolds. Is it possible that G is a Kähler group?

References

[A] D. Arapura, *Higgs line bundles, Green–Lazarsfeld sets, and maps of Kähler manifolds to curves*, Bull. Amer. Math. Soc. (1992).

[Ab] H. Abels, *An example of a finitely presented solvable group*, Homological group theory, 1977, London Math. Soc. Lecture Note Ser. **36**, 1979.

[ABR] D. Arapura, P. Bressler, and M. Ramachandran, *On the fundamental group of a compact Kähler manifold*, Duke. Math. J. (1993).

[AN] D. Arapura and M. Nori, *Solvable fundamental groups of algebraic varieties and Kähler manifolds* (to appear).

[Be] A. Beauville, *Surfaces algébriques complexes*, Astérisque **54** (1978).

[Bi] J. Birman, *Braids, Links and Mapping class groups*, Princeton Univ. Press, 1975.

[Bo] A. Borel, *Compact Clifford–Klein forms of symmetric spaces*, Topology (1963).

[Br] K. Brown, *Cohomology of groups*, Springer-Verlag, 1982.

[C] K. Corlette, *Nonabelian Hodge theory*, Proc. Sympos. Pure Math. **54** (1993).

[Ca] F. Campana, *Remarques sur les groupes de Kähler nilpotents*, Preprint (1993).

[CH] J. Carlson and L. Hernandez, *Harmonic maps from compact Kähler manifolds to exceptional hyperbolic spaces*, J. Geom. Analysis (1991).

[CT] J. Carlson and D. Toledo, *Harmonic mappings of Kähler manifolds to locally symmetric spaces*, Publ. Math. IHES (1989).

[CT2] ———, *Notes on nilpotent Kähler groups*, unpublished notes (1992).

[CT3] ———, *Quadratic presentations and nilpotent Kähler groups*, J. Geom. Analysis (to appear).

[CK] F. Catenese, J. Kollár, et al., *Trento examples*, Classification of irregular varieties, minimal models and Abelian varieties: Trento, 1990 (E. Ballico et al., ed.), Lecture Notes in Mathematics **1515**, Springer, 1992, pp. 133–139.

[De] P. Deligne, *Extensions central nonrésiduellement finis de groupes arithmetiques*, C. R. Acad. Sci. Paris Sér. I Math. **287** (1978).

[DGMS] P. Deligne, P. Griffiths, J. Morgan, and D. Sullivan, *Real homotopy theory of Kähler manifolds*, Invent. Math. (1975).

[G] M. Gromov, *Sur le groupe fondamental d'une variété kählerienne*, C. R. Acad. Sci. Paris Sér. I Math. (1988).

[GM] W. Goldman and J. Milson, *The deformation theory of representations of fundamental groups of compact Kähler manifolds*, Publ. Math. IHES (1988).

[GL] M. Green and R. Lazarsfeld, *Higher obstructions to deforming cohomology groups of line bundles*, J. Amer. Math. Soc. (1991).

[GS] M. Gromov and R. Schoen, *Harmonic maps into singular spaces and p-adic superrigidity for lattices in groups of rank one*, Publ. Math. IHES (1993).

[Hn] R. Hain, *Geometry of the mixed Hodge structure of the fundamental group of an algebraic variety*, Proc. Sympos. Pure Math., Amer. Math. Soc. (1987).

[JR1] F. Johnson and E. Rees, *On the fundamental group of a complex algebraic manifold*, Bull. London Math. Soc. (1987).

[JR2] ———, *The fundamental group of an algebraic variety*, Lecture Notes in Mathematics, vol. 1474, Springer-Verlag, 1991.

[GH] P. Griffiths and J. Harris, *Principles of algebraic geometry*, John Wiley, 1978.

[GM] P. Griffiths and J. Morgan, *Rational homotopy theory*, Birkhäuser, 1981.

[H] S. Helgason, *Differential geometry, Lie groups, and symmetric spaces*, Academic Press, 1978.

[KM] K. Kodaira and J. Morrow, *Complex manifolds*, Holt, Rinehart, Winston, 1971.

[Ko] T. Kohno, *Holonomy Lie algebras, logarithmic connections and lower central series of fundamental groups*, Contemp. Math. **90** (1989).

[L] R. Lyndon, *The cohomology of a group with a single defining relator*, Ann. Math (1950).

[M] J. Morgan, *The algebraic topology of smooth algebraic varieties*, Publ. Math. IHES (1978).

[Mi] J. Milnor, *Morse theory*, Princeton Univ. Press (1963).

[Q] D. Quillen, *Rational homotopy theory*, Ann. Math (1969).

[S] J.-P. Serre, *Sur la topologie des variétés algébriques en charactéristique p*, Symposium Internacional de Topologia Algebraica, Universidad Nacional Autonoma de Mexico, 1958, pp. 24–53.

[S1] C. Simpson, *Higgs bundles and local systems*, Publ. Math. IHES (1993).

[S2] ———, *The moduli space of rank local systems*, Ann. Sci. École Norm. Sup. (4) (1993).

[Sel] A. Selberg, *Discontinuous groups on higher dimensional symmetric spaces*, Contributions to function theory, Tata Inst. (1960).

[SV] A. Sommese and A. van de Ven, *Homotopy groups of pullbacks of varieties*, Nagoya Math. J. (1986).

[SW] G. Scott and C. Wall, *Topological methods in group theory*, London Math. Soc. Lecture Note Ser. **36** (1979).

[T1] D. Toledo, *Examples of fundamental groups of compact Kähler manifolds*, Bull. London Math. Soc. (1990).

[T2] ———, *Projective varieties with nonresidually finite fundamental group*, Publ. Math. IHES (1993).

[Ta] C. Taubes, *The existence of anti-selfdual conformal structures*, J. Differential Geom. (1992).

Donu Arapura
Department of Mathematics
Purdue University
West Lafayette IN 47907

E-mail address: dvb@math.purdue.edu

Complex Algebraic Geometry
MSRI Publications
Volume **28**, 1995

Vector Bundles on Curves and Generalized Theta Functions: Recent Results and Open Problems

ARNAUD BEAUVILLE

ABSTRACT. The moduli spaces of vector bundles on a compact Riemann surface carry a natural line bundle, the determinant bundle. The sections of this line bundle and its multiples constitute a non-abelian generalization of the classical theta functions. New ideas coming from mathematical physics have shed a new light on these spaces of sections—allowing notably to compute their dimension (Verlinde's formula). This survey paper is devoted to giving an overview of these ideas and of the most important recent results on the subject.

Introduction

It has been known essentially since Riemann that one can associate to any compact Riemann surface X an abelian variety, the Jacobian JX, together with a divisor Θ (well-defined up to translation) that can be defined both in a geometric way and as the zero locus of an explicit function, the Riemann theta function. The geometry of the pair (JX, Θ) is intricately (and beautifully) related to the geometry of X.

The idea that higher-rank vector bundles should provide a non-abelian analogue of the Jacobian appears already in the influential paper [We] of A. Weil (though the notion of vector bundle does not appear as such in that paper!). The construction of the moduli spaces was achieved in the 1960's, mainly by D. Mumford and the mathematicians of the Tata Institute. However it is only recently that the study of the determinant line bundles on these moduli spaces and of their spaces of sections has made clear the analogy with the Jacobian. This is

Partially supported by the European HCM project "Algebraic Geometry in Europe" (AGE), Contract CHRXCT 940557.

largely due to the intrusion of Conformal Field Theory, where these spaces have appeared (quite surprisingly for us!) as fundamental objects.

In these notes (based on a few lectures given in the Fall of 1992 at MSRI, UCLA and University of Utah), I will try to give an overview of these new ideas. I must warn the reader that this is by no means intended to be a complete account. I have mainly focused on the determinant line bundles and their spaces of sections, ignoring deliberately important areas like cohomology of the moduli spaces, moduli of Higgs bundles, relations with integrable systems, Langlands' geometric correspondence ... , simply because I felt it would have taken me too far afield. For the same reason I haven't even tried to explain why the mathematical physicists are so interested in these moduli spaces.

1. The moduli space $\mathcal{SU}_X(r)$

Let X be a compact Riemann surface of genus g. Recall that the Jacobian JX parametrizes line bundles of degree 0 on X. We will also consider the variety $J^{g-1}(X)$ that parametrizes line bundles of degree $g-1$ on X; it carries a *canonical* Theta divisor

$$\Theta = \{M \in J^{g-1}(X) \mid H^0(X, M) \neq 0\} .$$

For each line bundle L on X of degree $g-1$, the map $M \mapsto M \otimes L^{-1}$ induces an isomorphism of $(J^{g-1}(X), \Theta)$ onto (JX, Θ_L), where Θ_L is the divisor on JX defined by

$$\Theta_L = \{E \in JX \mid H^0(X, E \otimes L) \neq 0\} .$$

We know a great deal about the spaces $H^0(JX, \mathcal{O}(k\Theta))$. One of the key points is that the sections of $\mathcal{O}(k\Theta)$ can be identified with certain quasi-periodic functions on the universal cover of JX, the theta functions of order k. In this way one gets for instance that the dimension of $H^0(JX, \mathcal{O}(k\Theta))$ is k^g, that the linear system $|k\Theta|$ is base-point free for $k \geq 2$ and very ample for $k \geq 3$, and so on. One even obtains a rather precise description of the ring

$$\bigoplus_{k\geq 0} H^0(JX, \mathcal{O}(k\Theta)),$$

the graded ring of theta functions.

The character who will play the role of the Jacobian in these lectures is the moduli space $\mathcal{SU}_X(r)$ of (semistable) rank-r vector bundles on X with trivial determinant. It is an irreducible projective variety, whose points are isomorphism classes of vector bundles that are direct sums of stable vector bundles of degree 0 (a degree-0 vector bundle E is said to be *stable* if every proper subbundle of E has degree < 0). By the theorem of Narasimhan and Seshadri, the points of $\mathcal{SU}_X(r)$ are also the isomorphism classes of representations $\pi_1(X) \longrightarrow \mathrm{SU}(r)$ (hence the notation $\mathcal{SU}_X(r)$). The stable bundles form a smooth open subset of $\mathcal{SU}_X(r)$, whose complement—which parametrizes decomposable bundles—is singular (except in the cases $g \leq 1$ and $g = r = 2$, where the moduli space is smooth).

The reason for fixing the determinant is that the moduli space $\mathcal{U}_X(r)$ of vector bundles of rank r and degree 0 is, up to a finite étale covering, the product of $\mathcal{SU}_X(r)$ with JX, so the study of $\mathcal{U}_X(r)$ is essentially reduced to that of $\mathcal{SU}_X(r)$. Of course the moduli spaces $\mathcal{SU}_X(r, L)$ of semistable vector bundles with a fixed determinant $L \in \operatorname{Pic}(X)$ are also of interest; for simplicity, in these lectures I will concentrate on the most central case $L = \mathcal{O}_X$.

Observe that when $g \leq 1$ the spaces $\mathcal{SU}_X(r)$ consist only of direct sums of line bundles. Since these cases are quite easy to deal with directly, I will usually assume implicitly $g \geq 2$ in what follows.

2. The determinant bundle

The geometric definition of the theta divisor extends in a natural way to the higher-rank case. For any line bundle $L \in J^{g-1}(X)$, define

$$\Theta_L = \{E \in \mathcal{SU}_X(r) \mid h^0(X, E \otimes L) \geq 1\} .$$

This turns out to be a Cartier divisor on $\mathcal{SU}_X(r)$ [D-N] (the key point here is that the degrees are chosen so that $\chi(E \otimes L) = 0$). The associated line bundle $\mathcal{L} := \mathcal{O}(\Theta_L)$ does not depend on the choice of L. It is called the determinant bundle, and will play a central role in our story. It is in fact canonical, because of the following result (proved in [B1] for $r = 2$ and in [D-N] in general):

THEOREM 1. $\operatorname{Pic} \mathcal{SU}_X(r) = \mathbf{Z}\mathcal{L}$.

By analogy with the rank-one case, the global sections of the line bundles $\mathcal{L}^k$ are sometimes called *generalized theta functions*—we will briefly discuss this terminology in §10.

Can we describe $H^0(\mathcal{SU}_X(r), \mathcal{L})$ and the map $\varphi_{\mathcal{L}} : \mathcal{SU}_X(r) \dashrightarrow |\mathcal{L}|^*$ associated to $\mathcal{L}$? Let us observe that we can define a natural (rational) map of $\mathcal{SU}_X(r)$ to the linear system $|r\Theta|$, where Θ denotes the canonical Theta divisor on $J^{g-1}(X)$: for $E \in \mathcal{SU}_X(r)$, define

$$\theta(E) := \{L \in J^{g-1} \mid h^0(E \otimes L) \geq 1\} .$$

It is easy to see that $\theta(E)$ either is a divisor in $J^{g-1}(X)$ that belongs to the linear system $|r\Theta|$, or is equal to $J^{g-1}(X)$. This last case can unfortunately occur (see §3 below), but only for special E's, so we get a rational map

$$\theta : \mathcal{SU}_X(r) \dashrightarrow |r\Theta|.$$

THEOREM 2. *There is a canonical isomorphism*

$$H^0(\mathcal{SU}_X(r),\mathcal{L}) \xrightarrow{\sim} H^0(J^{g-1}(X),\mathcal{O}(r\Theta))^* \ ,$$

making the following diagram commutative:

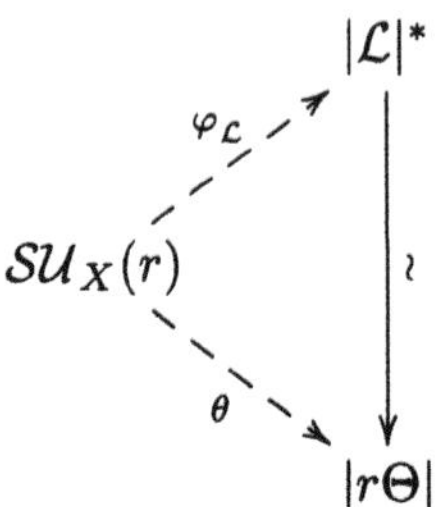

This is proved in [B1] for the rank-two case and in [B-N-R] in general. Let me say a few words about the proof. For L in $J^{g-1}(X)$ denote by H_L the hyperplane in $|r\Theta|$ consisting of divisors passing through L. One has $\theta^*H_L = \Theta_L$, so we get a linear map $\theta^* : H^0(J^{g-1}(X),\mathcal{O}(r\Theta))^* \longrightarrow H^0(\mathcal{SU}_X(r),\mathcal{L})$ whose transpose makes the above diagram commutative. It is easy to show that θ^* is injective, hence *the whole problem is to prove that* $\dim H^0(\mathcal{SU}_X(r),\mathcal{L}) = r^g$. This was done by constructing an r-to-one covering $\pi : Y \to X$ such that the pushforward map $\pi_* : JY \dashrightarrow \mathcal{SU}_X(r)$ is dominant, which gives an injective map of $H^0(\mathcal{SU}_X(r),\mathcal{L})$ into $H^0(JY,\mathcal{O}(r\Theta))$. Note that surjectivity of θ^* means that the linear system $|\mathcal{L}|$ is spanned by the divisors Θ_L for L in $J^{g-1}(X)$.

This theorem provides a relatively concrete description of the map $\varphi_{\mathcal{L}}$, and gives one some hope of being able to analyze the nature of this map—whether it is a morphism, an embedding, and so on. As we will see, this is a rather intriguing question, which is far from being completely understood. We first consider whether this map is everywhere defined or not.

3. Base points

It follows from Theorem 2 (more precisely, from the fact that the divisors Θ_L span the linear system $|\mathcal{L}|$) that the base points of $|\mathcal{L}|$ are the elements E of $\mathcal{SU}_X(r)$ such that $\theta(E) = J^{g-1}(X)$, that is, $H^0(E\otimes L)\neq 0$ for *all* line bundles L of degree $g-1$. The existence of such vector bundles was first observed by Raynaud [R]. Let me summarize his results in our language:

THEOREM 3. a) *For $r=2$, the linear system $|\mathcal{L}|$ has no base points.*

b) *For $r=3$, $|\mathcal{L}|$ has no base points if $g=2$, or if $g\geq 3$ and X is generic.*

c) *Let n be an integer ≥ 2 dividing g. For $r=n^g$, the system $|\mathcal{L}|$ has base points.*

In case (c), Raynaud's construction gives only finitely many base points. This leaves open a number of questions, which I will regroup under the same heading:

QUESTION 1. *Can one find more examples (say for other values of r)? Is the base locus of dimension > 0? On the opposite side, can one find a reasonable bound on the dimension of the base locus?*

Since the linear system $|\mathcal{L}|$ has (or may have) base points, we have to turn to its multiples. Here we have the following result of Le Potier [LP], improving an idea of [F1]:

PROPOSITION 1. *For $k > \frac{1}{4}r^3(g-1)$, the linear system $|\mathcal{L}^k|$ is base-point free.*

In fact Le Potier proves a slightly stronger statement: *given $E \in \mathcal{SU}_X(r)$ and $k > \frac{1}{4}r^3(g-1)$, there exists a vector bundle F on X of rank k and degree $k(g-1)$ such that $H^0(X, E \otimes F) = 0$* (in other words, what may fail with a line bundle always works with a rank-k vector bundle). Then

$$\Theta_F := \{E \in \mathcal{SU}_X(r) \mid H^0(X, E \otimes F) \neq 0\}$$

is a divisor of the linear system $|\mathcal{L}^k|$ which does not pass through E, hence the proposition.

The bound on k is certainly far from optimal; in view of Proposition 1 the most optimistic guess is

QUESTION 2. *Is $|\mathcal{L}^2|$ base-point free?*

Let me also mention the following question of Raynaud [R]:

QUESTION 3. *Given $E \in \mathcal{SU}_X(r)$, does there exist an étale covering $\pi : Y \to X$ such that $\theta(\pi^* E) \neq J^{g-1}(Y)$?*

4. Rank 2

The rank-two case is of course the simplest one; it has two special features. On one hand, by Theorem 3(a) (which is quite easy), we know that in this case $\varphi_{\mathcal{L}}$ is a *morphism*; we also know that this morphism is finite because $\mathcal{L}$ is ample. On the other hand, the linear system $|2\Theta|$ on $J^{g-1}(X)$ is particularly interesting because it contains the *Kummer variety* $\mathcal{K}_X$ of X. Recall that $\mathcal{K}_X$ is the quotient of the Jacobian JX by the involution $a \mapsto -a$, and that the map $a \mapsto \Theta_a + \Theta_{-a}$ of JX to $|2\Theta|$ (where as usual Θ_a denotes the translate of Θ by a) factors through an embedding $\kappa : \mathcal{K}_X \hookrightarrow |2\Theta|$. The non-stable part of $\mathcal{SU}_X(2)$ consists of vector bundles of the form $L \oplus L^{-1}$, for L in JX, and can therefore be identified with $\mathcal{K}_X$; recall that for $g \geq 3$ this is the singular locus of $\mathcal{SU}_X(2)$. Theorem 2 thus gives the following commutative diagram

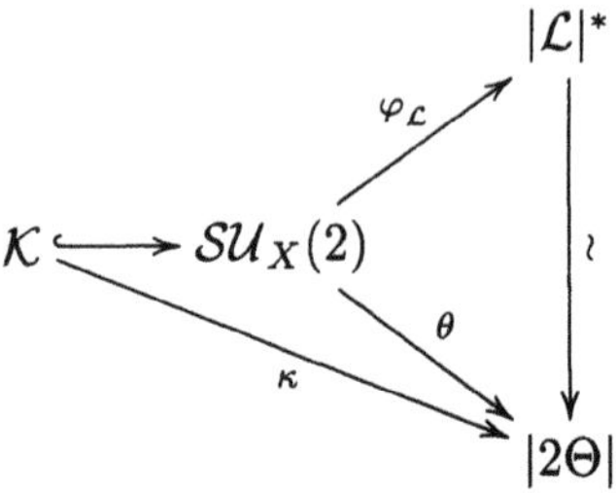

Let me summarize what is known about the structure of $\varphi_{\mathcal{L}}$ (or, what amounts to the same, of θ). Remember that the dimension of $|2\Theta|$ is $2^g - 1$.

THEOREM 4. a) *For $g = 2$, θ is an isomorphism of $\mathcal{SU}_X(2)$ onto $|2\Theta| \cong \mathbf{P}^3$* [N-R1].

b) *For $g \geq 3$, X hyperelliptic, θ is 2-to-1 onto a subvariety of $|2\Theta|$ that can be described in an explicit way* [D-R].

c) *For $g \geq 3$, X not hyperelliptic, θ is of degree one onto its image* [B1]. *Moreover if $g(X) = 3$ or if X is generic, θ is an embedding* ([N-R2]; [L], [B-V]).

The genus 3 (non hyperelliptic) case deserves a special mention: in this case Narasimhan and Ramanan prove that θ is an isomorphism of $\mathcal{SU}_X(2)$ onto a quartic hypersurface $\mathcal{Q}_4$ in $|2\Theta|$ ($\cong \mathbf{P}^7$). By the above remark, this quartic is singular along the Kummer variety $\mathcal{K}_X$. Now it had been observed a long time ago by Coble [C] that there exists a *unique* quartic hypersurface in $|2\Theta|$ passing doubly through $\mathcal{K}_X$! Therefore $\mathcal{Q}_4$ is nothing but Coble's hypersurface.

Part (c) of the theorem leaves open an obvious question:

QUESTION 4. *Is θ always an embedding for X non hyperelliptic?*

The case of a generic curve was proved first by Laszlo [L]; Brivio and Verra have recently developed a more geometric approach [B-V], which might hopefully lead to a complete answer to the question—though some serious technical difficulties remain at this moment.

Laszlo's method is to look at the canonical maps

$$\mu_k : S^k H^0(\mathcal{SU}_X(2), \mathcal{L}) \longrightarrow H^0(\mathcal{SU}_X(2), \mathcal{L}^k) .$$

Since we know that $\mathcal{L}^k$ is very ample for some (unknown!) integer k, surjectivity of μ_k for k large enough would imply that $\mathcal{L}$ itself is very ample. We can even dream of getting surjectivity for all k, which would mean that the image of $\mathcal{SU}_X(2)$ in $|2\Theta|$ is projectively normal. In [B2] the situation is completely analyzed for μ_2. Recall that a *vanishing thetanull* on X can be defined as a line bundle L on X with $L^{\otimes 2} \cong \omega_X$ and $h^0(L)$ even ≥ 2—this means that the corresponding theta function on JX vanishes at the origin, hence the name. Such a line bundle exists only on a special curve (more precisely on a divisor in the moduli space of curves). Then:

PROPOSITION 2. *If X has no vanishing thetanull, the map μ_2 is an isomorphism of $S^2H^0(\mathcal{SU}_X(2),\mathcal{L})$ onto $H^0(\mathcal{SU}_X(2),\mathcal{L}^2)$.*

More generally, if X has v vanishing thetanulls, one has

$$\dim \operatorname{Ker} \mu_2 = \dim \operatorname{Coker} \mu_2 = v.$$

This is only half encouraging (it shows that $\mathcal{SU}_X(2)$ is *not* projectively normal for curves with vanishing thetanulls), but note that the case $k = 2$ should be somehow the most difficult. On the positive side we have the following results:

PROPOSITION 3. a) *If X has no vanishing thetanull, the map*

$$\mu_4 : S^4H^0(\mathcal{SU}_X(2),\mathcal{L}) \longrightarrow H^0(\mathcal{SU}_X(2),\mathcal{L}^4)$$

is surjective [vG-P].

b) *If X is* generic, *the map $\mu_k : S^kH^0(\mathcal{SU}_X(2),\mathcal{L}) \longrightarrow H^0(\mathcal{SU}_X(2),\mathcal{L}^k)$ is surjective for k even and $\geq 2g-4$* [L].

As already mentioned, (b) implies that $\varphi_{\mathcal{L}}$ is an embedding for generic X.

5. The Verlinde formula

Trying to understand the maps μ_k raises inevitably the question of the dimension of the spaces $H^0(\mathcal{SU}_X(r),\mathcal{L}^k)$. We have seen that even the case $k=1$ is far from trivial—this is the essential part of [B-N-R]. So it came as a great surprise when the mathematical physicists claimed to have a general (and remarkable) formula for $\dim H^0(\mathcal{SU}_X(r),\mathcal{L}^k)$, called the Verlinde formula [V] (there is actually a more general formula for the moduli space of principal bundles under a semisimple group, but we will stick to the case of $\mathcal{SU}_X(r)$):

THEOREM 5.

$$\dim H^0(\mathcal{SU}_X(r),\mathcal{L}^k) = \Big(\frac{r}{r+k}\Big)^g \sum_{\substack{S\amalg T=[1,r+k]\\ |S|=r}} \prod_{\substack{s\in S\\ t\in T}} \left|2\sin\pi\frac{s-t}{r+k}\right|^{g-1}.$$

This form of the formula (shown to me by D. Zagier) is the simplest for an arbitrary rank; for small r or k I leave as a pleasant exercise to the reader to simplify it (hint: use $\prod_{p=1}^{n-1}(2\sin(p\pi/n)) = n$). One gets r^g in the case $k=1$, thus confirming Theorem 2, and in the rank-two case:

COROLLARY. $\dim H^0(\mathcal{SU}_X(2),\mathcal{L}^k) = (\frac{1}{2}k+1)^{g-1}\sum_{i=1}^{k+1}\frac{1}{(\sin(i\pi/(k+2)))^{2g-2}}\,.$

Note that the spaces $H^i(\mathcal{SU}_X(r),\mathcal{L}^k)$ vanish for $i>0$, by the Kodaira vanishing theorem (or rather its extension by Grauert and Riemenschneider), since the canonical bundle of $\mathcal{SU}_X(r)$ is equal to $\mathcal{L}^{-2r}$ [D-N]. Hence Theorem 5 gives actually $\chi(\mathcal{L}^k)$. The right-hand side must therefore be a polynomial in k, and take integral values, which is certainly not apparent from the formula! In fact I know no direct proof of these properties, except in the case $r=2$.

The leading coefficient of this polynomial is

$$(c_1(\mathcal{L})^n)/n!,$$

where $n=(r^2-1)(g-1)$ is the dimension of $\mathcal{SU}_X(r)$. This number, which is the volume of $\mathcal{SU}_X(r)$ for any Kähler metric with Kähler class $c_1(\mathcal{L})$, has been computed in a beautiful way by Witten [W1], using the properties of the Reidemeister torsion of a flat connection. The result is

$$\frac{c_1(\mathcal{L})^n}{n!} = r\,(2\pi)^{-2n}\mathrm{Vol}\,(\mathrm{SU}(r))^{2g-2}\sum_V\frac{1}{(\dim V)^{2g-2}}\,,$$

where V runs over all irreducible representations of $\mathrm{SU}(r)$, and the volume of $\mathrm{SU}(r)$ is computed with respect to a suitably normalized Haar measure. One should be able to deduce this formula from Theorem 5, but I don't know how to do that except in rank 2.

6. The Verlinde formula: finite-dimensional proofs

As soon as the Verlinde formula became known to mathematicians, it became a challenge for them to give a rigorous proof, so a wealth of proofs have appeared in the last few years. I will try to describe the ones I am aware of. The basic distinction is between the proofs using standard algebraic geometry, which up to now work only in the case $r=2$, and the proofs that use infinite-dimensional algebraic geometry to mimic the heuristic approach of the physicists—these work for all r. Let me start with the "finite-dimensional" proofs.

The first proof of this kind is due to Bertram and Szenes [B-S], who use the explicit description of the moduli space in the hyperelliptic case (cf. Theorem 4(b)) to compute $\chi(\mathcal{L}^k)$—which is the same for all smooth curves. (Actually they work with the moduli space $\mathcal{SU}_X(2,1)$ of vector bundles of rank 2 and fixed determinant of degree 1, which has the advantage of being smooth; they show that $\chi(\mathcal{L}^k)$ is the Euler–Poincaré characteristic of a certain vector bundle $\mathcal{E}_k$ on $\mathcal{SU}_X(2,1)$.)

A more instructive proof has been obtained by Thaddeus [T], building on ideas of Bertram and Bradlow–Daskalopoulos. The idea is to look at pairs consisting of a rank-two vector bundle E of fixed, sufficiently high degree, say $2d$, together with a nonzero section s of E. There is a notion of stability for these pairs—in fact there are various such notions, depending on an integer i with $0 \leq i \leq d$. For each of these values one gets a moduli space M_i, which is projective and smooth; the key point is that one passes from M_{i-1} to M_i (for $i \geq 2$) by a very simple procedure called a *flip*—blowing up a smooth subvariety and blowing down the exceptional divisor in another direction. Moreover, M_0 is just a projective space, M_1 is obtained by blowing up a smooth subvariety in M_0, while M_{d-1} maps surjectively to $\mathcal{SU}_X(2)$. In short one gets the following diagram

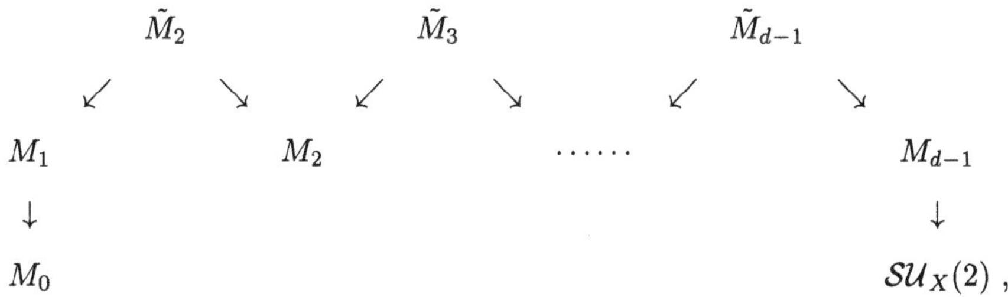

from which one deduces (with some highly non-trivial computations) the Verlinde formula.

Another completely different proof has been obtained by Zagier (unpublished). Not surprisingly, it is purely computational. Building on the work of Atiyah–Bott, Mumford and others, Zagier gives a complete description of the cohomology ring of $\mathcal{SU}_X(2,1)$; he is then able to write down explicitly the Riemann–Roch formula for $\chi(\mathcal{E}_k)$ (see above).

Another approach, due, I believe, to Donaldson and Witten, starts from Witten's formula for the volume of $\mathcal{SU}_X(r)$ (§5). More precisely, Witten gives also a formula for the volume of the moduli space of stable *parabolic* bundles; here again the stability depends on certain rational numbers, so we get a collection of volumes indexed by these rational numbers. It turns out that one can recover from these volumes all the coefficients of the polynomial $\chi(\mathcal{L}^k)$.

The last approach of this type I'd like to mention has been developed in [N-Rs] and [D-W1], and carried out successfully in [D-W2]. These authors attempt to prove directly that the spaces $H^0(\mathcal{SU}_X(2), \mathcal{L}^k)$ and the analogous spaces defined using parabolic vector bundles obey the so-called *factorization rules* (see below). Though this approach looks quite promising, the details are unfortunately quite technical, and an extension to higher rank seems out of reach.

7. The Verlinde formula: infinite-dimensional proofs

The idea here is to translate in algebro-geometric terms the methods of the physicists. Actually what the physicists are interested in is a vector space that plays a central role in Conformal Field Theory, the *space of conformal blocks* $B_X^k(r)$. This is defined as follows: let $\mathbf{C}((z))$ be the field of formal Laurent series in one variable. There is a canonical representation V_k of the Lie algebra $\mathfrak{sl}_r(\mathbf{C}((z)))$ (more precisely, of its universal central extension), called the *basic representation of level* k. Let $p \in X$; the affine algebra $A_X := \mathcal{O}(X - p)$ embeds into $\mathbf{C}((z))$ (by associating to a function its Laurent expansion at p). Then

$$B_X^k(r) := \{\ell \in V_k^* \mid \ell(Mv) = 0 \text{ for all} M \in \mathfrak{sl}_r(A_X) \text{ and } v \in V_k\} .$$

THEOREM 6. a) *There is a canonical isomorphism* $H^0(\mathcal{SU}_X(r), \mathcal{L}^k) \xrightarrow{\sim} B_X^k(r)$.

b) *The dimension of both spaces is given by the Verlinde formula* (given in Theorem 5).

There are by now several available proofs of these results. The fact that the dimension of $B_X^k(r)$ is given by the Verlinde formula follows from the work of Tsuchiya, Ueno and Yamada [T-U-Y]. They show that the dimension of $B_X^k(r)$ is independent of the curve X, *even if X is allowed to have double points*. Then it is not too difficult to express $B_X^k(r)$ in terms of analogous spaces for the normalization of X (this is called the *factorization rules* by the physicists). One is thus reduced to the genus-0 case (with marked points), that is, to a problem in the theory of representations of semisimple Lie algebras, which is non-trivial in general (actually I know no proof for the case of an arbitrary semisimple Lie algebra), but rather easy for the case of $\mathfrak{sl}_r(\mathbf{C})$.

Part (a) is proved (independently) in [B-L] and [F2]; actually Faltings proves both (a) and (b). He considers a smooth curve X degenerating to a stable curve X_s. It is not too difficult to show that $H^0(\mathcal{SU}_X(r), \mathcal{L}^k)$ embeds into $B_X^k(r)$, and that the $B_X^k(r)$'s are semicontinuous so that $\dim B_X^k(r) \le \dim B_{X_s}^k(r)$. Therefore the heart of [F2] is the proof of the inequality

$$\dim B_{X_s}^k(r) \le \dim H^0(\mathcal{SU}_X(r), \mathcal{L}^k) ,$$

of which I cannot say much, since I don't really understand it (note that the proof, as well as that of [T-U-Y], works in the more general set-up of principal bundles).

I would like to explain in a few words how we construct in [B-L] the isomorphism $H^0(\mathcal{SU}_X(r), \mathcal{L}^k) \xrightarrow{\sim} B_X^k(r)$, because I believe its importance goes far beyond the Verlinde formula. The basic object in the proof is not $\mathcal{SU}_X(r)$, but the *moduli stack* $\mathcal{SL}_X(r)$ parametrizing vector bundles E on X together with a trivialization of $\bigwedge^r E$. Though it appears at first glance as a rather frightening object, it is both more natural and easier to work with than the moduli space: basically, working with the moduli stack eliminates all the artificial problems of non-representability due to the fact that vector bundles have non-trivial automorphisms. The proof (which is entirely algebraic) has three steps:

1) We show that *the moduli stack $\mathcal{SL}_X(r)$ is isomorphic to the quotient stack* $\mathrm{SL}_r(A_X)\backslash \mathrm{SL}_r(\mathbf{C}((z)))/\mathrm{SL}_r(\mathbf{C}[[z]])$. The key point here is that a vector bundle with trivial determinant is *algebraically trivial* over $X - p$. (Hint: show that such a bundle has always a nowhere vanishing section, and use induction on the rank). We choose a small disk $D \subset X$ around p (actually, to avoid convergence problems we take $D = \mathrm{Spec}(\mathcal{O})$, where $\mathcal{O}$ is the completed local ring of X at p, but this makes essentially no difference). We then consider triples (E, ρ, σ), where E is a vector bundle on X, ρ an *algebraic* trivialization of E over $X - p$ and σ a trivialization of E over D. Over $D - p$ these two trivializations differ by a holomorphic map $D - p \longrightarrow \mathrm{GL}_r(\mathbf{C})$ that is meromorphic at p, that is, given by a Laurent series $\gamma \in \mathrm{GL}_r(\mathbf{C}((z)))$. Conversely, given such a matrix γ, one can use it to glue together the trivial bundles on $X - p$ and D and recover the triple (E, ρ, σ). Since we want γ in $\mathrm{SL}_r(\mathbf{C}((z)))$ we impose moreover that $\bigwedge^r \rho$ and $\bigwedge^r \sigma$ coincide over $D - p$. This gives a bijection of the set of triples (E, ρ, σ) (up to isomorphism) onto $\mathrm{SL}_r(\mathbf{C}((z)))$.

To get rid of the the trivializations, we have to mod out by the automorphism group of the trivial bundle over D and $X - p$. We get the following diagram:

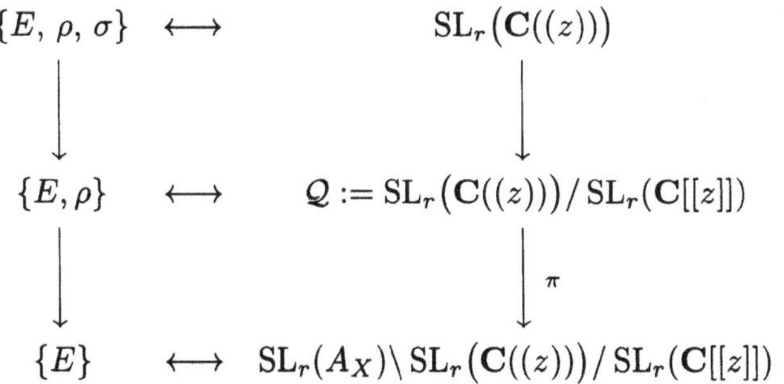

$$\begin{array}{ccc}
\{E,\rho,\sigma\} & \longleftrightarrow & \mathrm{SL}_r(\mathbf{C}((z))) \\
\downarrow & & \downarrow \\
\{E,\rho\} & \longleftrightarrow & \mathcal{Q} := \mathrm{SL}_r(\mathbf{C}((z)))/\mathrm{SL}_r(\mathbf{C}[[z]]) \\
\downarrow & & \downarrow \pi \\
\{E\} & \longleftrightarrow & \mathrm{SL}_r(A_X)\backslash \mathrm{SL}_r(\mathbf{C}((z)))/\mathrm{SL}_r(\mathbf{C}[[z]])\ .
\end{array}$$

Of course I have only constructed a bijection between the set of isomorphism classes of vector bundles on X with trivial determinant and the set of double classes $\mathrm{SL}_r(A_X)\backslash \mathrm{SL}_r(\mathbf{C}((z)))/\mathrm{SL}_r(\mathbf{C}[[z]])$; with some technical work one shows that the construction actually gives an isomorphism of stacks.

2) Recall that, if $Q = G/H$ is a homogeneous space, one associates to any character $\chi : H \to \mathbf{C}^*$ a line bundle L_χ on Q: it is the quotient of the trivial bundle $G \times \mathbf{C}$ on G by the action of H defined by $h(g, \lambda) = (gh, \chi(h)\lambda)$. We apply this to the homogeneous space $\mathcal{Q} = \mathrm{SL}_r(\mathbf{C}((z)))/\mathrm{SL}_r(\mathbf{C}[[z]])$ (this is actually an

ind-variety, i.e., the direct limit of an increasing sequence of projective varieties). By (1) we have a quotient map $\pi : \mathcal{Q} \longrightarrow \mathcal{SL}_X(r)$. The line bundle $\pi^*\mathcal{L}$ does *not* admit an action of $\mathrm{SL}_r\big(\mathbf{C}((z))\big)$, but of a group $\widehat{\mathrm{SL}}_r\big(\mathbf{C}((z))\big)$ that is a central $\mathbf{C}^*$-extension of $\mathrm{SL}_r\big(\mathbf{C}((z))\big)$. This extension splits over the subgroup $\mathrm{SL}_r(\mathbf{C}[[z]])$, so that $\mathcal{Q}$ is isomorphic to $\widehat{\mathrm{SL}}_r\big(\mathbf{C}((z))\big)/(\mathbf{C}^* \times \mathrm{SL}_r(\mathbf{C}[[z]]))$. Then $\pi^*\mathcal{L}$ is the line bundle L_χ, where $\chi : \mathbf{C}^* \times \mathrm{SL}_r(\mathbf{C}[[z]]) \longrightarrow \mathbf{C}^*$ is the first projection.

3) A theorem of Kumar and Mathieu provides an isomorphism $H^0(\mathcal{Q}, L_\chi^k) \cong V_k^*$. From this and the definition of a quotient stack one can identify

$$H^0(\mathcal{SL}_X(r), \mathcal{L}^k)$$

with the subspace of V_k^* invariant under $\mathrm{SL}_r(A_X)$. This turns out to coincide with the subspace of V_k^* invariant under the Lie algebra $\mathfrak{sl}_r(A_X)$, which is by definition $B_X^k(r)$. Finally a Hartogs type argument gives $H^0(\mathcal{SU}_X(r), \mathcal{L}^k) \cong H^0(\mathcal{SL}_X(r), \mathcal{L}^k)$.

To conclude let me observe that all the proofs I have mentioned are rather indirect, in the sense that they involve either degeneration arguments or sophisticated computations. The simplicity of the formula itself suggests the following question:

QUESTION 5. *Can one find a direct proof of Theorem 5?*

What I have in mind is for instance a computation of $\chi(\mathcal{L}^k)$ by simply applying the Riemann–Roch formula; this requires the knowledge of the Chern numbers of the moduli space. In [W2], Witten proposes some very general conjectures that should give the required Chern numbers for $\mathcal{SU}_X(r)$: the preprint [S] sketches how the Verlinde formula follows from these conjectures. Jeffrey and Kirwan have proved some of the Witten's conjectures, and I understand that they are very close to a proof of the Verlinde formula along these lines.

8. The strange duality

Let me denote by $\mathcal{U}_X^*(k)$ the moduli space of semistable vector bundles of rank k and degree $k(g-1)$ on X; it is isomorphic (non canonically) to $\mathcal{U}_X(k)$. A special feature of this moduli space is that it carries a *canonical* theta divisor Θ_k: set-theoretically one has

$$\Theta_k = \{E \in \mathcal{U}_X^*(k) \mid H^0(X, E) \neq 0\} .$$

Put $\mathcal{M} := \mathcal{O}(\Theta_k)$. Consider the morphism $\tau_{k,r} : \mathcal{SU}_X(r) \times \mathcal{U}_X^*(k) \longrightarrow \mathcal{U}_X^*(kr)$ defined by $\tau_{k,r}(E,F) = E \otimes F$. An easy application of the theorem of the square shows that $\tau_{k,r}^*(\mathcal{O}(\Theta_{kr}))$ is isomorphic to $pr_1^*\mathcal{L}^k \otimes pr_2^*\mathcal{M}^r$. Now $\tau_{k,r}^*\Theta_{kr}$ is the divisor of a section of this line bundle, well-defined up to a scalar; by the Künneth theorem we get a linear map $\vartheta_{k,r} : H^0(\mathcal{U}_X^*(k), \mathcal{M}^r)^* \longrightarrow H^0(\mathcal{SU}_X(r), \mathcal{L}^k)$, well-defined up to a scalar. In this section I want to discuss the following conjecture:

QUESTION 6. CONJECTURE: *The map $\vartheta_{k,r}$ is an isomorphism.*

I heard of this statement three or four years ago, as being well-known to the physicists. The conjecture is discussed at length, and extended to vector bundles of arbitrary degree, in [D-T].

Let me discuss a few arguments in favor of the conjecture.

a) The case $k = 1$ is exactly Theorem 2.
b) The two spaces have the same dimension. To prove this one needs to compute the dimension of $H^0(\mathcal{U}_X^*(k), \mathcal{M}^r)$; this is easy (assuming the Verlinde formula!) because the map $\tau_{1,k} : \mathcal{SU}_X(k) \times J^{g-1}(X) \longrightarrow \mathcal{U}_X^*(k)$ is an étale (Galois) covering of degree k^{2g}, and $\tau_{1,k}^*(\mathcal{M}) \cong pr_1^*\mathcal{L} \otimes pr_2^*\mathcal{O}_J(k\Theta)$. Therefore we get
$$\dim H^0((\mathcal{U}_X^*(k), \mathcal{M}^r) = \chi(\mathcal{M}^r) = \frac{1}{k^{2g}}\ \chi(\mathcal{L}^r)\ \chi(\mathcal{O}_J(kr\Theta))$$
$$= \frac{r^g}{k^g}\ \dim H^0((\mathcal{SU}_X(k), \mathcal{L}^r)\ .$$
Now, Theorem 5 shows that $k^{-g} \dim H^0((\mathcal{SU}_X(k), \mathcal{L}^r)$ is symmetric in k and r, which proves our assertion.
c) Therefore it is enough to prove, for example, the surjectivity of the map $\vartheta_{k,r}$, which has the following geometric meaning:

QUESTION 6′. *The linear system $|\mathcal{L}^k|$ in $\mathcal{SU}_X(r)$ is spanned by the divisors Θ_F, for F in $\mathcal{U}_X^*(k)$.*

(Recall from §3 that Θ_F is the locus of vector bundles $E \in \mathcal{SU}_X(r)$ such that $H^0(X, E \otimes F) \neq 0$). As an application, taking vector bundles F of the form $L_1 \oplus \cdots \oplus L_k$ with $L_i \in J^{g-1}(X)$, one deduces from Proposition 3(b) that *the conjecture holds for $r = 2$ and k even and* $\geq 2g - 4$ (in this way we get the result for a generic curve only, but using the methods in §9 below I can extend it to every curve).

9. The projective connection

So far we have considered the moduli space $\mathcal{SU}_X(r)$ for a fixed curve X. What can we say about the vector spaces $H^0(\mathcal{SU}_X(r), \mathcal{L}^k)$ when the curve X is allowed to vary? It is again a remarkable discovery of the mathematical physicists that these vector spaces are *essentially independent of the curve*. To explain this in mathematical terms, consider a family of (smooth) curves $(X_t)_{t \in T}$, parametrized by a variety T; for t in T, let us denote by $\mathcal{L}_t$ the determinant line bundle on $\mathcal{SU}_{X_t}(r)$. Then:

THEOREM 7. *The linear systems $|\mathcal{L}_t^k|$ define a* **flat** *projective bundle over T.*

Here again we have by now a number of proofs for this result. The first mathematical proof is due to Hitchin [H], following the method used by Welters in the rank-one case; I understand that Beilinson and Kazhdan had a similar proof (unpublished). A different approach, inspired by the work of the physicists, appears in [F1]. Finally, one of the main ingredients in [T-U-Y] is the construction of a flat vector bundle over T whose fibre at $t \in T$ is the space of conformal blocks $B_{X_t}^k(r)$ (the curves of the family are required to have a marked point $p_t \in X_t$, together with a distinguished tangent vector $v_t \in T_{p_t}(X_t)$). Thanks to Theorem 6 this provides still another construction of our flat projective bundle. I have no doubt that all these constructions give the same object, but I must confess that I haven't checked it.

Let us take for T the moduli space $\mathcal{M}_g$ of curves of genus g (here again, the correct object to consider is the moduli stack, but let me ignore this). We get a flat projective bundle over $\mathcal{M}_g$, which corresponds to a projective representation

$$\rho_{r,k} : \Gamma_g \longrightarrow \mathrm{PGL}\big(H^0(\mathcal{SU}_X(r), \mathcal{L}^k)\big)$$

of the fundamental group $\Gamma_g = \pi_1(\mathcal{M}_g, \mathrm{X})$. This group, called the *modular group* by the physicists and the *mapping class group* by the topologists, is a fundamental object: it carries all the topology of $\mathcal{M}_g$. So a natural question is

QUESTION 7. *What is the representation $\rho_{r,k}$?*

This is a rather intriguing question. In the rank-one case, the analogue of $H^0(\mathcal{SU}_X(r), \mathcal{L}^k)$ is the space V_k of k-th order theta functions; the group Γ_g acts on V_k through its quotient $\mathrm{Sp}\,(2g, \mathbf{Z})$ (take this with a grain of salt when k is odd), and this action is explicitly described by the classical "transformation formula" for theta functions—which shows in particular that the action factors through a finite quotient of $\mathrm{Sp}\,(2g, \mathbf{Z})$. Using Theorem 2 we get an analogous description for arbitrary rank in the case $k = 1$. I expect that the general case is far more complicated, and in particular that $\rho_{r,k}$ doesn't factor through $\mathrm{Sp}\,(2g, \mathbf{Z})$, but I know no concrete example where this happens.

Conformal Field Theory predicts that the connection should be (projectively) *unitary*. This is one of the remaining challenges for mathematicians:

QUESTION 8. *Find a flat hermitian metric H_X on $H^0(\mathcal{SU}_X(r), \mathcal{L}^k)$* (i.e., one such that the image of $\rho_{r,k}$ is contained in $PU(H_X)$).

Here again the rank-one case is well-known, and Theorem 2 gives an answer in the case $k = 1$—though not a very explicit one.

10. Are there generalized theta functions?

I believe that the above results give some evidence that the spaces

$$H^0(\mathcal{SU}_X(r), \mathcal{L}^k)$$

are non-abelian analogues of the spaces $H^0(JX, \mathcal{O}(k\Theta))$. There is however one aspect of the picture that is missing so far in higher rank, namely the analytic description of the sections of $\mathcal{O}(k\Theta)$ as holomorphic functions. Clearly the theory of theta functions cannot be extended in a straightforward way, if only because $\mathcal{SU}_X(r)$ is simply connected.

One possible approach is provided by our description of the moduli stack as a double quotient $\mathrm{SL}_r(A_X)\backslash \mathrm{SL}_r(\mathbf{C}((z)))/\mathrm{SL}_r(\mathbf{C}[[z]])$ (§7). The pullback of the determinant line bundle $\mathcal{L}$ to the group $\widehat{\mathrm{SL}}_r(\mathbf{C}((z)))$, which is a $\mathbf{C}^*$-extension of $\mathrm{SL}_r(\mathbf{C}((z)))$, is trivial, so we should be able to express sections of $\mathcal{L}^k$ as functions on $\widehat{\mathrm{SL}}_r(\mathbf{C}((z)))$. This is done in [B-L] in one particular case: we prove that the pullback of the divisor $\Theta_{(g-1)p}$ is the divisor of a certain algebraic function on $\widehat{\mathrm{SL}}_r(\mathbf{C}((z)))$ known as the τ function. (This idea appears already in [Br], with a slightly different language.) However, it is not clear how to express other sections of $\mathcal{L}$ (and even less of $\mathcal{L}^k$) in a similar way.

There are other possible ways of describing sections of $\mathcal{L}^k$ by holomorphic functions. One of these, explored by D. Bennequin, is to pull back $\mathcal{L}$ to $\mathrm{SL}_r(\mathbf{C})^g$ by the dominant map $\mathrm{SL}_r(\mathbf{C})^g \dashrightarrow \mathcal{SU}_X(r)$ that maps a g-tuple $(M_1, \ldots, M_g)$ to the flat vector bundle $E(\rho_M) \in \mathcal{SU}_X(r)$ associated to the representation $\rho_M : \pi_1(X) \longrightarrow \mathrm{SL}_r(\mathbf{C})$ with $\rho(a_i) = I$, $\rho(b_j) = M_j$ (here $(a_1, \ldots, a_g, b_1, \ldots, b_g)$ are the standard generators of $\pi_1(X)$).

Despite these attempts I am afraid we are still far of having a satisfactory theory of generalized theta functions. So I will end up this survey with a loosely formulated question:

QUESTION 9. *Is there a sufficiently simple and flexible way of expressing the elements of $H^0(\mathcal{SU}_X(r), \mathcal{L}^k)$ as holomorphic functions?*

REFERENCES

[B1] A. Beauville, *Fibrés de rang 2 sur les courbes, fibré déterminant et fonctions thêta*, Bull. Soc. Math. France **116** (1988), 431–448.

[B2] ———, *Fibrés de rang 2 sur les courbes, fibré déterminant et fonctions thêta II*, Bull. Soc. Math. France **119** (1991), 259–291.

[B-L] A. Beauville and Y. Laszlo, *Conformal blocks and generalized theta functions*, Comm. Math. Phys. **164** (1994), 385–419.

[B-N-R] A. Beauville, M. S. Narasimhan and S. Ramanan, *Spectral curves and the generalized theta divisor*, J. Reine Agnew. Math. **398** (1989), 169–179.

[B-S] A. Bertram and A. Szenes, *Hilbert polynomials of moduli spaces of rank* 2 *vector bundles II*, Topology **32** (1993), 599–609.

[B-V] S. Brivio and A. Verra, *The Theta divisor of $\mathcal{SU}_C(2)$ is very ample if C is not hyperelliptic and Noether-Lefschetz general*, preprint (1993).

[Br] J. -L. Brylinski, *Loop groups and non-commutative theta-functions*, preprint (1989).

[C] A. Coble, *Algebraic geometry and theta functions*, Amer. Math. Soc. Colloq. Publ., vol. 10, Providence, 1929, third edition, (1969).

[D-W1] G. Daskalopoulos and R. Wentworth, *Local degenerations of the moduli space of vector bundles and factorization of rank* 2 *theta functions* I, Math. Ann. **297** (1993), 417–466.

[D-W2] ———, *Factorization of rank* 2 *theta functions II: proof of the Verlinde formula*, preprint (1994).

[D-R] U. V. Desale and S. Ramanan, *Classification of vector bundles of rank* 2 *on hyperelliptic curves*, Invent. Math. **38** (1976), 161–185.

[D-T] R. Donagi and L. Tu, *Theta functions for* SL(n) *versus* GL(n), Math. Res. Let. 1 (1994), 345–357.

[D-N] J. M. Drezet and M. S. Narasimhan, *Groupe de Picard des variétés de modules de fibrés semi-stables sur les courbes algébriques*, Invent. Math. **97** (1989), 53–94.

[F1] G. Faltings, *Stable G-bundles and projective connections*, J. Algebraic Geom. **2** (1993), 507–568.

[F2] ———, *A proof for the Verlinde formula*, J. Algebraic Geom. **3** (1994), 347–374.

[H] N. Hitchin, *Flat connections and geometric quantization*, Comm. Math. Phys. **131** (1990), 347–380.

[L] Y. Laszlo, *A propos de l'espace des modules des fibrés de rang* 2 *sur une courbe*, Math. Annalen **299** (1994), 597–608.

[N-R1] M. S. Narasimhan and S. Ramanan, *Moduli of vector bundles on a compact Riemann surface*, Ann. of Math. **89** (1969), 19–51.

[N-R2] ———, *2θ-linear systems on Abelian varieties*, Vector bundles on algebraic varieties, Oxford University Press, 1987, pp. 415–427.

[N-Rs] M. S. Narasimhan and T. R. Ramadas, *Factorization of generalized Theta functions I*, Invent. Math. **114** (1993), 565–623.

[LP] J. Le Potier, *Module des fibrés semi-stables et fonctions thêta*, preprint (1993).

[R] M. Raynaud, *Sections des fibrés vectoriels sur une courbe*, Bull. Soc. Math. France **110** (1982), 103–125.

[S] A. Szenes, *The combinatorics of the Verlinde formula*, preprint (1994).

[T] M. Thaddeus, *Stable pairs, linear systems and the Verlinde formula*, Invent. Math. **117** (1994), 317–353.

[T-U-Y] A. Tsuchiya, K. Ueno, and Y. Yamada, *Conformal field theory on universal family of stable curves with gauge symmetries*, Adv. Stud. Pure Math. **19** (1989), 459–566.

[V] E. Verlinde, *Fusion rules and modular transformations in 2d conformal field theory*, Nuclear Phys. B **300** (1988), 360-376.

[vG-P] B. van Geemen and E. Previato, *Prym varieties and the Verlinde formula*, Math. Annalen **294** (1992), 741–754.

[We] A. Weil, *Généralisation des fonctions abéliennes*, J. de Math. P. et App. **17** (1938), no. 9ème sér., 47–87.

[W1] E. Witten, *On quantum gauge theories in two dimensions*, Comm. Math. Phys. **141** (1991), 153–209.

[W2] ———, *Two-dimensional gauge theories revisited*, J. Geom. Phys. **9** (1992), 308–368.

Arnaud Beauville
URA 752 du CNRS
Mathématiques – Bât. 425
Université Paris-Sud
91 405 Orsay Cedex, France

E-mail address: arnaud@matups.matups.fr

Complex Algebraic Geometry
MSRI Publications
Volume **28**, 1995

Recent Results in Higher-Dimensional Birational Geometry

ALESSIO CORTI

ABSTRACT. This note surveys some recent results on higher-dimensional birational geometry, summarising the views expressed at the conference held at MSRI in November 1992. The topics reviewed include semistable flips, birational theory of Mori fiber spaces, the logarithmic abundance theorem, and effective base point freeness.

Contents

1. Introduction

The purpose of this note is to survey some recent results in higher-dimensional birational geometry. A glance to the table of contents may give the reader some idea of the topics that will be treated. I have attempted to give an informal presentation of the main ideas, emphasizing the common grounds, addressing a general audience. In §3, I could not resist discussing some details that perhaps only the expert will care about, but hopefully will also introduce the non-expert reader to a subtle subject.

Perhaps the most significant trend in Mori theory today is the increasing use, more or less explicit, of the logarithmic theory. Let me take this opportunity

This work at the Mathematical Sciences Research Institute was supported in part by NSF grant DMS 9022140.

to advertise the Utah book [Ko], which contains all the recent software on log minimal models. Our notation is taken from there.

I have kept the bibliography to a minimum and made no attempt to give proper credit for many results. The reader who wishes to know more about the results or their history could start from the references listed here and the literature quoted in those references.

The end of a proof or the absence of it will be denoted with a $\square$. Anyway here proof always means "proof": a bare outline will be given at best, usually only a brief account of some of what the author considers to be the main ideas.

In preparing the manuscript, I received considerable help from J. Kollár, J. McKernan and S. Mori. The responsibility for mistakes is of course entirely mine.

2. Notation, Minimal Models, etc.

The aim of this section is to introduce the basic notation and terminology to be used extensively in the rest of the paper, and to give a quick reminder of minimal model theory, including the logarithmic theory.

Unless otherwise explicitly declared, we shall work with complex projective normal varieties.

An *integral Weil divisor* on a variety X is a formal linear combination $B = \sum b_i B_i$ with integer coefficients of irreducible subvarieties $B_i \subset X$ of codimension 1. A $\mathbb{Q}$-*Weil divisor* is a linear combination with rational coefficients. We say B is *effective* if all $b_i \geq 0$. We denote with $\lceil B \rceil = \sum \lceil b_i \rceil B_i$ the round-up and with $\lfloor B \rfloor = \sum \lfloor b_i \rfloor B_i$ the round-down of B.

A *Cartier divisor* is a line bundle together with the divisor of a meromorphic section. A $\mathbb{Q}$-Weil divisor B is $\mathbb{Q}$-*Cartier* if mB is Cartier (i.e., the divisor of the meromorphic section of some line bundle) for some integer $m > 0$. Numerical equivalence of $\mathbb{Q}$-Cartier divisors is defined below and is denoted by $\equiv$, while linear equivalence of Weil divisors is denoted by $\sim$.

A Cartier divisor D on X is *nef* if $D \cdot C \geq 0$ for all algebraic curves $C \subset X$.

A Weil divisor is *qef* (quasieffective) if it is a limit of effective $\mathbb{Q}$-divisors. Clearly a nef divisor is qef.

Here are some elementary examples with nef and qef. Let X be a smooth algebraic surface containing no -1 curves, and let K_X be the the canonical class. If K_X is qef, it is also nef; on the other hand, if K_X is not qef, adjunction terminates (that is, $|D + mK| = \varnothing$ for all D, where m is a sufficiently large positive integer) and X must be uniruled (the reader who does not find these assertions obvious is invited to prove them now as an exercise). It is also true, but it appears to be more delicate to show, that if K_X is qef, (a multiple of) K_X is actually effective (and even free from base points).

A variety X is *$\mathbb{Q}$-factorial* if every Weil divisor on X is $\mathbb{Q}$-Cartier. This property is local in the Zariski but not in the analytic topology, which makes this notion quite subtle and may lead to confusion. However, to avoid sometimes serious technical problems, when running the minimal model algorithm starting with a variety X, we shall always assume that X is $\mathbb{Q}$-factorial.

Let $f : X \dashrightarrow Y$ be a birational map, $B \subset X$ a Weil divisor. Then $f_*B \subset Y$ denotes the *birational transform*. It is the Weil divisor on Y defined as follows. Let $U \subset X$ be an open subset of X with $\operatorname{codim}_X(X \setminus U) \geq 2$ and $f_U : U \to Y$ a morphism representing f. Then f_*B is the (Zariski) closure in Y of $f_{U*}B$. If $B = \sum b_i B_i$ with B_i prime, $f_{U*}B = \sum b_i f_{U*}B_i$, where by definition $f_{U*}B_i = f_U(B_i)$ if $f_U(B_i)$ is a divisor, and $f_{U*}B_i = 0$ otherwise.

Consistently, if $B \subset Y$ is a Weil divisor on Y, we use $f_*^{-1}B$ for $(f^{-1})_*B$. This way we don't confuse it with the set theoretic preimage $f^{-1}(B)$, defined when f is a morphism, or the pullback as a ($\mathbb{Q}$-)Cartier divisor f^*B, which makes sense when f is a morphism and B happens to be a ($\mathbb{Q}$-)Cartier divisor.

I will now describe the minimal model algorithm. The starting point is always a normal projective $\mathbb{Q}$-factorial variety X, together with a $\mathbb{Q}$-Cartier ($\mathbb{Q}$-)divisor D. All or part of D may only be defined up to linear equivalence of Weil divisors (such is the case in the most important example, where $D = K_X$). In practice, some additional conditions are imposed on the pair (X, D) in order for the program to work: for instance, $D = K_X + B$, where B is an effective $\mathbb{Q}$-Weil divisor and (X, B) is log canonical (see Definition (2.1) below).

In the theory of Zariski decomposition, one tries to remove from D the "negative" part, thus writing $D = D' + D''$, where D'' is nef and $H^0(mD'') = H^0(mD)$ for all positive integers m. Instead, we modify X by a sequence of birational operations, removing all configurations in X, where D is negative. We inductively construct a sequence of birational maps $X = X_0 \dashrightarrow X_1 \dashrightarrow \cdots \dashrightarrow X_n = X'$, and divisors D_i on X_i, in such a way that $H^0(X_{i-1}, mD_{i-1}) = H^0(X_i, mD_i)$ for all i and all positive integers m, and $D_n = D'$ is nef. After introducing the cone of curves with some motivating remarks, I will describe in more detail how this is done.

Let $NS(X) \otimes \mathbb{R} \subset H^2(X, \mathbb{R})$ be the real Néron–Severi group (it is the subgroup of $H^2(X, \mathbb{R})$ generated by the classes of Cartier divisors). Numerical equivalence on $NS(X) \otimes \mathbb{R}$ is defined by setting $D \equiv D'$ if and only if $D \cdot [C] = D' \cdot [C]$ for all algebraic curves $C \subset X$, and we let $N^1(X)$ be the quotient of $NS(X) \otimes \mathbb{R}$ by $\equiv$ (if X has rational singularities, $\equiv$ is trivial on $NS(X) \otimes \mathbb{Q}$). We also define $N_1(X)$ to be the dual real vector space. By definition $N_1(X)$ can be naturally identified with the free real vector space generated by all algebraic curves $C \subset X$, modulo the obvious notion of numerical equivalence. The *Kleiman–Mori* cone $\overline{NE}(X)$ is by definition the closure (in the natural Euclidean topology) of the convex cone $NE(X) \subset N_1(X)$ generated by the (classes of) algebraic curves.

Let X be a normal projective variety, $f : X \to Y$ a projective morphism. If H is any ample divisor on Y and $C \subset X$ a curve, $f^*H \cdot C = 0$ if and only if C is contained in a fiber of f. By the Kleiman criterion for ampleness, this shows that $\{[C] \mid f(C) = pt\}$ generates a *face* $F \subset \overline{NE}(X)$. A one-dimensional face of $\overline{NE}(X)$ is called an *extremal ray*.

Here is how Mori theory works. Start with $(X_0, D_0) = (X, D)$. Assume that the chain $X_0 \overset{t_0}{\dashrightarrow} \cdots X_{k-1} \overset{t_{k-1}}{\dashrightarrow} X_k$ and divisors D_k on X_k have been constructed. If D_k is nef on X_k, we have reached a *D-minimal model*, and the program stops here at $(X', D') = (X_k, D_k)$. Otherwise if D_k is not nef we need to show that $\overline{NE}(X_k)$ is locally finitely generated in $\{z \mid D_k \cdot z < 0\}$ (*cone theorem*), pick an extremal ray $R \subset \overline{NE}(X_k)$ with $D_k \cdot R < 0$, and construct a morphism $f : X_k \to Y$ to a normal variety Y, with the property that a curve $C \subset X$ is contracted by f if and only if $[C] \in R$ (*contraction theorem*). There are three possibilities for f:

1) $\dim(Y) < \dim(X_k)$. In this case we say that X_k, together with the fibration $f : X_k \to Y$, is a *D-Mori fiber space*. The program stops here, and we are happy because we have a strong structural description of the final product.

2) f is birational and the exceptional set of f contains a divisor (it is easy to see that the exceptional set must then consist of a single prime divisor E). In this case we say that f is a *divisorial contraction* and set $t_k = f$, $X_{k+1} = Y$, $D_{k+1} = f_* D_k$ and proceed inductively.

3) f is birational and *small* (or *flipping*)—that is, the exceptional set of f does not contain a divisor. The problem here is that $f_* D_k$ is not $\mathbb{Q}$-Cartier, so it makes no sense to ask whether it is nef. The appropriate modification $t_k : X_k \dashrightarrow X_{k+1}$ is an entirely new type of birational transformation, called to the *D-opposite* or *D-flip* of f, and $D_{k+1} = t_{k*} D_k$. The flip $f' : X_{k+1} \to Y$ is characterized by the following properties: D_{k+1} is ($\mathbb{Q}$-Cartier and) f'-nef, and f' is small.

If every step of the program (cone theorem, contraction theorem and flip theorem) can be shown to exist, and there is no infinite sequence of flips (i.e., *termination of flips* holds), the minimal model algorithm is established, and terminates in a minimal model or a Mori fibration. Presently, this has only been completed in the following cases:

a) $\dim(X) \leq 3$, $D = K_X$, and X has *terminal* or *canonical* singularities.

b) $\dim(X) \leq 3$, $D = K_X + B$, and the pair (X, B) is *log terminal* or *log canonical*.

c) X is a toric variety and $D = K_X$.

Let me recall the definition:

(2.1) DEFINITION. Let X be a variety, $B = \sum b_i B_i$ a Weil divisor with B_i prime divisors and $0 < b_i \leq 1$ (we allow $B = \varnothing$). We say that the pair (X, B) is terminal (resp. canonical, resp. log terminal, resp. log canonical), or that the divisor $K + B$ is terminal (resp. canonical, etc.) if $K_X + B$ is $\mathbb{Q}$-Cartier and for all normal varieties Y and birational morphism $f : Y \to X$ with exceptional prime divisors E_i we have:

$$K_Y + f_*^{-1}B = f^*(K_X + B) + \sum a_i E_i$$

with all $a_i > 0$ (resp. $a_i \geq 0$, resp. $a_i > -1$, resp. $a_i \geq -1$).

The numbers a_i in definition (2.1) only depend on the valuations ν_i of $\mathbb{C}(X)$ associated to E_i, and can be computed on any normal Z with birational $Z \to X$ such that ν_i is a divisor in Z. This way we may define the discrepancy $a(\nu, K_X + B)$ for any (algebraic) valuation ν with small center on X. In this language $K_X + B$ is log terminal if and only if $a(\nu, K_X + B) > -1$ for all ν with small center on X, and likewise for the other adjectives. The following result is very easy to prove but crucial in all questions concerning termination:

(2.2) THEOREM. *Let X be a normal variety, and $D = K_X + B$. Let $t : X \dashrightarrow X'$ be a step (divisorial contraction or flip) in the D-minimal model program. Then $a(\nu, K_{X'} + B') \geq a(\nu, K_X + B)$ for all valuations ν with small center in X. Also, $a(\nu, K_{X'} + B') > a(\nu, K_X + B)$ if and only if t is not an isomorphism at the center of ν in X. In particular, if $K_X + B$ is log terminal (terminal, canonical, log canonical), so is $K_{X'} + B'$.*

PROOF. The result is an exercise in the definition if t is a divisorial contraction. If t is a flip, see [Ko, 2.28]. □

I wish to spend a few words on the meaning of the D-minimal model program.

First, when X is smooth and $D = K_X$, this is the genuine minimal model program or Mori program: for surfaces, it consists in locating and contracting -1 curves until none can be found.

When X is smooth, B is a (reduced) divisor on X with global normal crossings, and $D = K_X + B$, the D-minimal model program is a sort of Mori program for the open variety $U = X \setminus B$. The (birational) category of open varieties was introduced by Iitaka, and it may be argued that it too is a primitive God-given entity.

For D general, the D-minimal model program should be considered as a way to obtain some kind of "Zariski decomposition" of D. In particular, it is not clear (certainly not to me) a priori that there should be any interesting reason at all to consider general logarithmic divisors $K_X + B$, where $B = \sum b_i B_i$ is allowed to have rational coefficients b_i, $0 < b_i \leq 1$. Nevertheless, these divisors have been profitably used (especially by Kawamata) since the earlier days of the theory, to direct or construct portions of the (genuine) Mori program, especially in relation to flops and flips. They are crucial in Shokurov's approach to the flipping problem [Ko]. Also, as I will try to show in this survey, the general log category is a key tool in several recent advances in higher-dimensional birational geometry.

3. Semistable Flips

This section summarizes some of the results of [Ka] and [S].

The starting point is a projective semistable family of surfaces $f : X \to T$ over a Dedekind scheme T. This means that f is a projective morphism, X is a regular scheme of dimension 3, and all fibers of f are reduced divisors with global normal crossings. The goal is to establish the minimal model algorithm over T:

$$X/T \dashrightarrow X_1/T \dashrightarrow \cdots \dashrightarrow X_i/T \dashrightarrow X_{i+1}/T \dashrightarrow \cdots \dashrightarrow X_N/T = Y/T.$$

The idea is as follows. It may be assumed that T is the spectrum of a DVR; let $S \subset X$ be the central fiber. By assumption $K_X + S$ is log terminal and $K_X + S \equiv K_X$, since S is a fiber so numerically trivial over T. Every step in the K-minimal model program is then a step in the $(K + S)$-minimal model program. In particular, by (2.2), each $K_{X_i} + S_i$ is log terminal. This implies that K_{S_i} too is log terminal, that is, S_i has quotient singularities. The presence of S_i gives a good control of singularities appearing on X_i, so flips are easier to construct than in the general case. The most important message in [Ka] is that a good understanding of the log surface S is all that is needed to establish the minimal model program in this particular setting. Semistable families of surfaces is the simplest situation where a proof of the existence of flips is known. It is important to revisit this proof in the light of the most recent advances in Mori theory, in the hope to gain a better understanding of flips in general.

[Ka] extends the known results to all Dedekind schemes T (with a new construction of flips), and [S] gives a new proof over $\mathbb{C}$, thereby giving an explicit classification of singularities appearing on each intermediate $S_i \subset X_i$ (and hence on the final product Y/T) which, among other things, allows to explain an old result of Kulikov.

From now on, for simplicity of notation, T will be the spectrum of a DVR $\mathcal{O}$ with parameter τ, $f : X \to T$ a projective morphism from a 3-fold, $S \subset X$ the fiber over the closed point. Starting with (3.4), $T = \Delta \subset \mathbb{C}$ is a small disc centered at the origin.

The approach in [Ka] is based on the following classification.

(3.1) LEMMA. *Let T be the spectrum of a complete DVR $\mathcal{O}$ with uniformizing parameter τ and algebraically closed residue field $k = \mathcal{O}/(\tau)$. Let $f : X \to T$ be a* *morphism from a three-dimensional scheme X. Assume that f is smooth outside $S = f^*(0)$, and that $K_X + S$ is log terminal (the conditions imply that X has terminal singularities and S is reduced). If T has positive or mixed characteristic, assume moreover that X is Cohen–Macaulay (this is automatic in characteristic 0). Let $x \in S$ be a point, r the index of K_X at x. One of the following alternatives holds, describing the completion $\hat{\mathcal{O}}_{X,x}$ as a $\mathcal{O}$-algebra:*

(3.1.1) $\hat{\mathcal{O}}_{X,x} \cong \mathcal{O}[[x, y, z]]/(xyz - \tau)$,

(3.1.2) $\hat{\mathcal{O}}_{X,x} \cong \mathcal{O}[[x, y, z]]^{\mathbb{Z}_r}/(xy - \tau)$, *where $\mathbb{Z}_r$ acts with weights $(a, -a, 1)$ for some $(a, r) = 1$.*

(3.1.3) $r > 1$ *and* $\hat{\mathcal{O}}_{X,x} \cong \mathcal{O}[[x, y, z]]^{\mathbb{Z}_r}/(xy - F(z^r))$, *where $\mathbb{Z}_r$ acts with weights $(a, -a, 1)$ for some $(a, r) = 1$, or*
$r = 1$ *and* $\hat{\mathcal{O}}_{X,x} \cong \mathcal{O}[[x, y, z]]/(G(x, y, z))$.

Let $\overline{F} \in k[z]$ (resp. $\overline{G} \in k[[x, y, z]]$) be the reduction of F (resp. G) mod τ. Then $xy - \overline{F}$ (resp. $\overline{G}$) defines a rational double point, which must then be of type A_{hr} for some h (resp. could be any rational double point).

PROOF. The proof goes by analyzing the canonical cover $\pi : (x \in X') \to (x \in X)$. There are two technical problems if $\mathcal{O}$ has positive or mixed characteristic:

a) It is important to know that $S \subset X$ (hence $S' \subset X'$) is S_2. Here we need the assumption that X is Cohen–Macaulay. In characteristic 0 it is known that log terminal singularities are Cohen–Macaulay, but the proof relies on the Grauert–Riemenschneider vanishing theorem, which in general does not hold in positive or mixed characteristic. It is not known, in positive or mixed characteristic, whether log terminal singularities are Cohen–Macaulay in general.

b) The canonical cover depends on the choice of an isomorphism $\mathcal{O}_X \to \omega_X^{[r]}$. A poor choice may produce a nonnormal X'. □

(3.2) CONSTRUCTION. As a consequence of the classification (3.1), one can (partially) resolve singular points $x \in X$ of index $r > 1$ with weighted blow-ups of the ambient toric variety, in a way that will be now described.

First we introduce some terminology. Let $x \in S \subset X$ be a *higher-index* ($r > 1$) singular point of the type described in (3.1.3). Then $x \in S$ is a rational double point of type A_{hr}, and we call h the *complexity* of x (we also say that x is a *simple* point if $h = 1$, or if we are in case (3.1.2)). Also, the integer $n(x) = \operatorname{ord}_\tau F(0)$ is called the *axial multiplicity* of x ($n(x) = 1$ always in case (3.1.2)).

(3.2.1) With the notation of (3.1), if $r > 1$, let $f : Z \to X$ be induced by the weighted blow-up of $\mathcal{O}[x, y, z]$ with weights of (x, y, z, τ) equal to $\frac{1}{r}(a, r-a, 1, r)$.

Let $S' = f^*S$. The following can be checked directly, by a straightforward, if tedious, calculation in toric geometry:

(3.2.1.1) Z is normal and Cohen–Macaulay. Also, $K_Z = f^*K_X + \frac{1}{r}E$, where E is the exceptional divisor. This means that f "extracts" valuation(s) over x with minimal discrepancy.

(3.2.1.2) Assume that $h(x) = 1$, that is, $x \in X$ is a simple point. Then Z has terminal singularities, $K_Z + S'$ is log terminal, and E is irreducible. More precisely, Z has three singular points $z_i \in Z$, and the completed local rings are as follows:

$\hat{\mathcal{O}}_{Z,z_1} \cong \mathcal{O}[[x,y,z]]^{\mathbb{Z}_a}/(xy-\tau)$, with weights $(r,-r,1)$,
$\hat{\mathcal{O}}_{Z,z_2} \cong \mathcal{O}[[x,y,z]]^{\mathbb{Z}_{r-a}}/(xy-\tau)$, with weights $(r,-r,1)$,
$\hat{\mathcal{O}}_{Z,z_3} \cong \mathcal{O}[[x,y,z]]^{\mathbb{Z}_r}/(xy-\frac{1}{\tau}F(\tau z^r))$, where $\mathbb{Z}_r$ acts with weights $(a,-a,1)$.

In particular z_1 and z_2 have index $< r$, and z_3 has index r but axial multiplicity $n(z) = n(x) - 1$. It can thus be reasonably asserted that Z has simpler singularities than X.

If $h(x) > 1$, various elements change the picture just given for the case $h = 1$, all of which are slightly disturbing for the purpose of constructing flips with the method of [Ka]. Specifically, $K_Z + S'$ is not log terminal (but very mildly so), Z does not always have terminal singularities (it always has a cA_{s-2} curve, for $s = \min\{h(x), n(x)\}$), and the exceptional set E has two irreducible components when $n(x) \geq 2$.

(3.2.2) Assume $h = h(x) \geq 2$; so we are in case (3.1.3). Let $m = \operatorname{ord} F$ (if $F = \sum_{j\geq 0} F_j z^{rj}$; by definition $m = \min\{\operatorname{ord}_\tau(F_j) + j \mid j \geq 0\}$). Let $f : Z \to X$ be induced by the weighted blow-up of $\mathcal{O}[x,y,z]$ with weights of (x,y,z,τ) equal to $(i+\frac{a}{r}, m-i-\frac{a}{r}, \frac{1}{r}, 1)$, for any $0 \leq i < m$. Also write, as above, $S' = f^*S$. Then as above Z is normal, Cohen–Macaulay, and $K_Z = f^*K_X + \frac{1}{r}E$. Again $K_Z + S'$ is not log terminal (very mildly so), but at least now Z has terminal singularities and E is irreducible. More precisely, Z has four singular points $z_i \in Z$, and the completed local rings are as follows:

$\hat{\mathcal{O}}_{Z,z_1} \cong \mathcal{O}[[x,y,z]]^{\mathbb{Z}_{ri+a}}/(xy-\tau)$, with weights $(r,-r,1)$,
$\hat{\mathcal{O}}_{Z,z_2} \cong \mathcal{O}[[x,y,z]]^{\mathbb{Z}_{r(m-i)-a}}/(xy-\tau)$, with weights $(r,-r,1)$,
$\hat{\mathcal{O}}_{Z,z_3} \cong \mathcal{O}[[x,y,z]]^{\mathbb{Z}_r}/(xy-\frac{1}{\tau^m}F(\tau z^r))$, where $\mathbb{Z}_r$ acts with weights $(a,-a,1)$,
$z_4 \in Z$ is a cA_l terminal singular point (for some l). It is at this point that $K_Z + S'$ is not log terminal.

To summarize (3.2), one can use the classification of singularities and weighted blow-ups to construct partial resolutions $f : Z \to X$ (perhaps using (3.2.2) if $h > 1$). The blow-up constructed in (3.2.1), in the case $h = 1$, is useful for constructing flips inductively, because (a) singularities on Z have lower index or axial multiplicity, (b) $K_Z + S'$ is again log terminal. The main disadvantage of

(3.2.2) is that, although the singularities $z_i \in Z$ do appear to be "simpler" than $x \in X$, this is difficult to make precise because $z_1 \in Z$ or $z_2 \in Z$ has index $> r$.

Next I state the main result in [Ka]:

(3.3) THEOREM. *Let X/T be a semistable family of surfaces over a Dedekind scheme T. Assume that the chain*

$$X \overset{t_1}{\dashrightarrow} \cdots \dashrightarrow X_{i-1} \overset{t_i}{\dashrightarrow} X_i$$

has been constructed, where each $t_j : X_{j-1} \dashrightarrow X_j$ is the modification (divisorial contraction or flip) associated to an extremal ray $R_{j-1} \subset \overline{NE}(X_{j-1}/T)$ with $K_{X_{j-1}} \cdot R_{j-1} < 0$. Then the cone theorem, on the structure of $\overline{NE}(X_i/T)$, the contraction theorem and the flip theorem hold for X_i/T in the usual way. In particular, if K_{X_i} is not nef over T, there is an extremal contraction $\varphi : X_i \to Z$. If φ is divisorial, set $X_{i+1} = Z$; if φ is small, let $t_{i+1} : X_i \dashrightarrow X_{i+1}$ be the flip. This inductively establishes the minimal model program for X/T.

PROOF. The cone and contraction theorem for X_i are deduced from the cone and contraction theorem for the log surface S_i. Because of this, and because we need the classification (3.1) to establish the existence of flips, we need at each step to check that X_j is Cohen–Macaulay.

Existence of flips goes as follows. Let $\varphi : X_i \to W$ be a flipping contraction. As usual I may assume that the exceptional set consists of a single irreducible curve $C \cong \mathbb{P}^1 \subset X_i$. It can be seen that there is a point $x \in C \subset X_i$, where K_{X_i} has index $r > 1$. Choose x with maximal r.

If $x \in X_i$ looks like (3.1.3), assume moreover that $h = 1$. Let $f : Z \to X_i$ be the blow-up at $x \in X_i$ described in (3.2.1), and $C' \subset Z$ the proper transform of C. A direct calculation taking place on S_i, and involving the classification of surface quotient singularities, shows that $K_Z \cdot C' \leq 0$. Then C' can be flipped or flopped on Z over W. A sequence of flips on Z/W (possibly preceded by a single flop), followed by a divisorial contraction, gives the original flip of $X_i \to W$. Flips on Z exist because Z is simpler (3.2.1.2). The flop at the beginning sometimes occurs, but in this situation it can be easily constructed directly, so no previous knowledge of (terminal) flops is assumed here.

This point deserves to be emphasized a little more. The reader of [Ka] may notice that, as a byproduct of the above mentioned calculation leading to the inequality $K_Z \cdot C' \leq 0$, it is easy to construct a divisor B in a neighbourhood of $C \subset X$ such that $K_X + B$ is log terminal and numerically trivial on C (we do not even need $h = 1$ for this). Traditionally, one is done at this point via a covering trick: if U is a double covering of a neighbourhood of C with branch divisor B, the flip of X is a quotient of the flop of U. Here, however, U has canonical singularities, so we need to know that canonical flops exist for this approach to work. The whole purpose of using the toric blow-up $f : Z \to X$

(with the complications if $h > 1$ soon to be described) is to avoid using canonical flops at this stage (in fact, any knowledge of flops at all).

If $x \in X_i$ looks like (3.1.3) and $h > 1$, a base change followed by a small analytic $\mathbb{Q}$-factorialization reduces to the case $h = 1$. $\square$

In working with the singularities appearing on X_i/T in (3.3), two attitudes are possible. The maximalistic view of [Ka] studies $S_i \subset X_i$ such that $K_{X_i} + S_i$ is log terminal. In the minimalistic view of [S], one tries to understand exactly which $S_i \subset X_i$ can originate from a semistable X/T. This is based on the following observations.

Assume the germ $(x \in S \subset X)$ is analytically isomorphic to the germ $0 \in (xy + z^2 = 0) \subset \mathbb{C}^3$. Although $K + S$ is log terminal and S is Cartier, this germ does not appear on a semistable family of surfaces. If $f : Y \to X$ is an embedded resolution of S, then $f^*S \subset Y$ is not reduced, nor, consequently, semistable (one has to base extend in order to construct a semistable reduction of $S \subset X$).

Now consider $(x \in S \subset X) \cong (0 \in (z = w) \subset (xy + zw = 0))$. Here $S = f^*(0)$ for $f = z - w$, and this singularity certainly appears on the minimal model of a semistable family of surfaces. However $x \in X$ can be resolved with a single small blow-up, also resolving the double point $x \in S$.

This suggests that Du Val singularities should not appear on special fibers of analytic $\mathbb{Q}$-factorializations of minimal models of semistable families of surfaces. Note that if $X \to \Delta$ is projective and $X' \to X$ is an analytic $\mathbb{Q}$-factorialization of X, the composite $X' \to \Delta$ is often not projective.

(3.4) DEFINITION. Let $f : X \to \Delta$ be a not necessarily projective morphism from a complex threefold X to a small disk $\Delta \subset \mathbb{C}$, and let $S = f^*(0)$ be the central fiber. Then f is S-semistable (Shokurov-semistable) if X has terminal singularities and there is a resolution $g : Y \to X$ such that $(fg)^*(0)$ is a global normal crossing divisor.

The above definition has the obvious disadvantage that the conditions are difficult to check. In particular it is not clear at all (but true) that if $X \to \Delta$ is S-semistable and $X/\Delta \dashrightarrow X'/\Delta$ is a flip, then $X' \to \Delta$ is also S-semistable. The next two statements summarize the main results of [S]:

(3.5) THEOREM. *Let $f : X \to \Delta$ be S-semistable and projective, and*

$$X/\Delta \dashrightarrow \cdots \dashrightarrow X_i/\Delta \dashrightarrow X_{i+1}/\Delta \dashrightarrow \cdots \dashrightarrow X_N/\Delta$$

a minimal model program for X over Δ. Then each $X_i \to \Delta$ (and so also the final product $X_N \to \Delta$) is S-semistable. $\square$

The following result should be compared with (3.1) above, and represents the shift in philosophy from [Ka] to [S] (the main difference is (3.6.3)):

(3.6) THEOREM. *Let $X \to \Delta$ be S-semistable, and as usual let $S = f^*(0)$ be the central fiber. Let $x \in S \subset X$ be a point, r the index of K_X at x. One of the following holds:*

(3.6.1) *x belongs to exactly three irreducible components of S and the germ $x \in S \subset X$ is analytically isomorphic to*

$$0 \in (xyz = 0) \subset \mathbb{C}^3.$$

(3.6.2) *x belongs to exactly two irreducible components of S and the germ $x \in S \subset X$ is analytically isomorphic to*

$$0 \in (xy = 0) \subset \mathbb{C}^3/\mathbb{Z}_r,$$

where $\mathbb{Z}_r$ acts with weights $(a, -a, 1)$ and $(a, r) = 1$.

(3.6.3) *x belongs to exactly one irreducible component of S. Let $X' \to X$ be a small analytic $\mathbb{Q}$-factorialization of X, $S' \subset X'$ the proper transform, $x_i \in S'$ a singular point lying over x. The germ $x_i \in S' \subset X'$ is analytically isomorphic to*

$$0 \in (t = 0) \subset (xy + z^r + t^{n_i} = 0 \subset \mathbb{C}^4/\mathbb{Z}_r)$$

where $n_i > 0$ is an integer (this is the axial multiplicity (3.2)), $\mathbb{Z}_r$ acts on $\mathbb{C}^4$ with weights $(a, -a, 1, 0)$, and $(a, r) = 1$. □

In particular, as a consequence of (3.6), S-semistable analytically $\mathbb{Q}$-factorial singularities have a very simple structure.

In [S], the existence of flips and the above results are proved at the same time, by induction on the depth of S-semistable singularities. The *depth* of a *S*-semistable singularity, depth(x, S, X), is by definition the minimum number of g-exceptional divisors in a resolution $g : Y \to X$ as in (3.4). The minimum is to be taken among all g's admitting a factorization $g = g_N \circ g_{N-1} \circ \cdots \circ g_1$ in $g_i : Y_i \to Y_{i+1}$ (here $Y = Y_1$ and $X = Y_{N+1}$) such that g_i is projective *locally analytically over* Y_{i+1}. This approach seems to have two main technical nuisances, which I try to briefly describe. On the one hand the minimal model algorithm only works for projective morphisms, so we are not free to analytically $\mathbb{Q}$-factorialize everything. On the other hand, the inductive approach of Shokurov uses certain divisorial extractions that only exist in the analytic category. As a result, one is led to run the minimal model program on varieties that are not necessarily $\mathbb{Q}$-factorial. A typical paradox arising in this context is that sometimes it is necessary to "flip" divisorial contractions.

As a corollary one immediately obtains an old result of Kulikov:

(3.7) Corollary. *Let $X \to \Delta$ be a semistable degeneration of* K3 *surfaces, $X/\Delta \dashrightarrow Y/\Delta$ the minimal model, $Y' \to Y$ a small analytic $\mathbb{Q}$-factorialization, $Y' \to \Delta$ the induced morphism. Then $Y' \to \Delta$ is semistable (in the usual sense).*

Proof. The generic fiber Y_η is a minimal K3 surface, so $K_{Y_\eta} \sim 0$. Because K_X is nef over Δ, we have $K_X \equiv 0$. Since the central fiber is reduced, $K_X \sim 0$, and so K_X has index 1 at all singular points. The result now follows instantly from (3.6). □

Note again that $Y' \to \Delta$ in (3.7) is often not projective.

4. Birational theory of Mori fibrations

This section summarizes the results of [C] on birational maps between Mori fibrations in dimension three.

Recall that X admits a Mori fibration if X has $\mathbb{Q}$-factorial terminal singularities and there is a morphism $\varphi : X \to S$ to a lower-dimensional normal S, with $-K_X$ φ-ample and $\rho(X) - \rho(S) = 1$. The condition $\rho(X) - \rho(S) = 1$ is very important and means that a class D on X is the pull-back of a class on S whenever $D \cdot C = 0$ for some curve C contained in a fiber of φ.

Philosophically, the statements arise from the systematic attempt to consider the morphism φ as a built in structure, not merely as an accessory of the variety X.

We work with the following incarnation of the log category. Let X be a variety with $\mathbb{Q}$-factorial terminal singularities, $\mathcal{H}$ a linear system without base divisors on X. The pair $(X, b\mathcal{H})$ is terminal (canonical, etc.) if and only if for all normal varieties Y and birational morphism $f : Y \to X$ with exceptional divisors E_i one has:

$$K_Y + bf_*^{-1}\mathcal{H} = f^*(K_X + b\mathcal{H}) + \sum a_i E_i$$

with all $a_i > 0$ ($a_i \geq 0$, etc.). Here $f_*^{-1}\mathcal{H}$ denotes the linear system without fixed divisors induced on Y by $\mathcal{H}$, and is called the birational transform.

Throughout this section, a birational map Φ and the following notation will be fixed:

$$\begin{array}{ccc} X & \overset{\Phi}{\dashrightarrow} & X' \\ \downarrow{\scriptstyle\varphi} & & \downarrow{\scriptstyle\varphi'} \\ S & & S' \end{array}$$

Choose a sufficiently large (and divisible) positive integer μ', and an extremely ample divisor A' on S'. Then the linear system $\mathcal{H}' = |-\mu' K_{X'} + \varphi'^* A'|$ is very ample on X'.

Let $\mathcal{H} = \Phi_*^{-1}\mathcal{H}'$ be the birational transform on X. For some positive rational number μ and (not necessarily ample) class A on S, we have $\mathcal{H} \equiv -\mu K_X + \varphi^* A$

(recall that $\rho(X) - \rho(S) = 1!$).

Let $p : (Y, \mathcal{H}_Y) \to (X, \mathcal{H})$ be a resolution of X and the base locus of $\mathcal{H}$.

The situation is summarized in the following diagram:

$$\begin{array}{ccccccc} & & & Y & & & \mathcal{H}_Y = q^*\mathcal{H}' \\ & & \swarrow^{p} & & \searrow^{q} & & \\ \mathcal{H} \equiv -\mu K_X + \varphi^* A & & X & \overset{\Phi}{\dashrightarrow} & X' & & \mathcal{H}' = \mid -\mu' K_{X'} + \varphi'^* A' \mid \\ & & \downarrow{\scriptstyle \varphi} & & \downarrow{\scriptstyle \varphi'} & & \\ & & S & & S' & & \end{array}$$

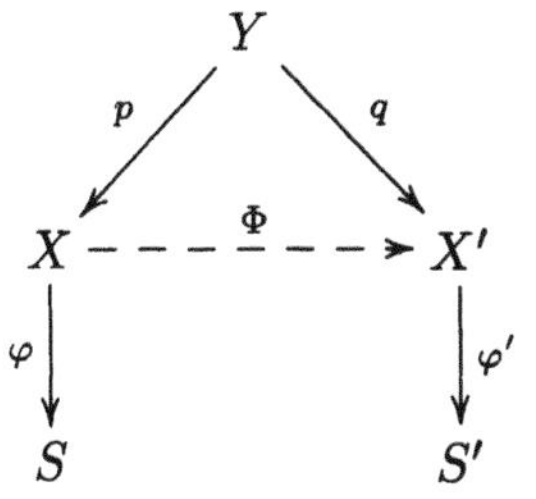

The next result is the key to understand the numerical geometry of Φ. Recall that a class in $H^2_{\mathbb{R}}$ is *qef* (quasieffective) if it is a limit of classes of effective $\mathbb{Q}$-divisors.

(4.1) THEOREM (NÖTHER–FANO INEQUALITIES).

(4.1.1) $\mu \geq \mu'$, *and equality holds only if* Φ *induces a rational map* $S \dashrightarrow S'$ *and* Φ^{-1} *is contracting.*

(4.1.2) *If* $K_X + \frac{1}{\mu}\mathcal{H}$ *is canonical and qef,* Φ *is an isomorphism in codimension one, and it induces an isomorphism* $X_\eta \cong X'_{\eta'}$ *of generic fibers. In particular,* $\mu = \mu'$.

(4.1.3) *If* $K_X + \frac{1}{\mu}\mathcal{H}$ *is canonical and nef,* Φ *is an isomorphism, and it also induces an isomorphism* $S \cong S'$. *In particular,* $\mu = \mu'$.

PROOF. Let's prove (4.1.1) first. We have

$$K_Y + \frac{1}{\mu'}\mathcal{H}_Y = q^*(K_{X'} + \frac{1}{\mu'}\mathcal{H}') + \sum a'_i E_i = q^*\varphi'^* A' + \sum a'_i E_i,$$

with all $a'_i > 0$ because X' has terminal singularities. Also

$$K_Y + \frac{1}{\mu'}\mathcal{H}_Y = p^*(K_X + \frac{1}{\mu'}\mathcal{H}) + \sum a_i E_i,$$

where I know nothing of a_i. Let $C \subset X$ be a general curve contained in a fiber of φ, let $C' \subset X'$ be the transform in X' and C'' the transform in Y. Then

$$(K_X + \frac{1}{\mu'}\mathcal{H}) \cdot C = (K_Y + \frac{1}{\mu'}\mathcal{H}_Y) \cdot C'' = \varphi'^* A' \cdot C' + \sum a'_i E_i \cdot C'' \geq 0.$$

Thus $\mu \geq \mu'$, and $\mu = \mu'$ only if C' is contained in a fiber of φ' and every q-exceptional divisor is also p-exceptional, i.e., Φ^{-1} is contracting.

To prove (4.1.2), reverse the above argument. This gives $\mu = \mu'$. Then (4.1.1) and some extra work imply that Φ is an isomorphism in codimension one.

One has to work a little harder to get (4.1.3). This is standard once the statement of what we wish to prove is known. □

The intent is to think of μ as the "degree" of Φ. The plan is to define a class of elementary maps, and use (4.1) to construct an elementary map $\Psi : X/S \dashrightarrow X_1/S_1$ such that $\Phi_1 = \Psi^{-1} \circ \Phi : X_1/S_1 \dashrightarrow X/S$ has degree $\mu_1 < \mu$. Note that the definition itself of μ makes crucial use of the fibration $X \to S$ and the property $\rho(X) - \rho(S) = 1$, while the proof of (4.1) also uses the similar structure of X'. To emphasize even more these concepts, I observe that the elementary $\Psi : X/S \dashrightarrow X_1/S_1$ might very well be the identity map of X, only changing the Mori fibration structure $X \to S$ to $X \to S_1$ (see links of type IV below). Indeed, I win in this game if I can reduce the degree μ. For instance, in the surface case, if $X \to S \cong \mathbb{P}^1 \times \mathbb{P}^1 \overset{\pi_1}{\to} \mathbb{P}^1$ is the first projection, the degree μ equals $\frac{1}{2}\mathcal{H} \cdot \pi_1^{-1}(*)$, and it might well be that the degree is smaller when measured with respect to the second projection $\mathbb{P}^1 \times \mathbb{P}^1 \overset{\pi_2}{\to} \mathbb{P}^1$.

The following is the most important statement of [C]. It is due to Sarkisov, as is the philosophy of its proof.

(4.2) THEOREM. *Let $\phi : X/S \dashrightarrow X'/S'$ be a birational map between Mori fibrations, with $\dim(X) \leq 3$. There is a factorization $\Phi = \Phi_n \cdots \Phi_1$:*

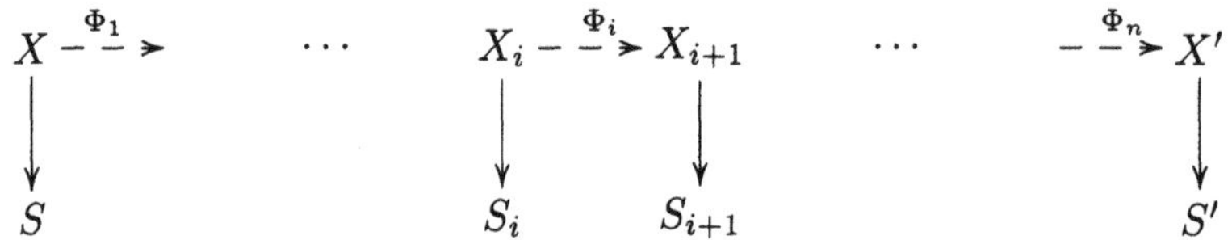

where $\Phi_i : X_i/S_i \dashrightarrow X_{i+1}/S_{i+1}$ is one of the following elementary maps:

(4.2.1) Links of type I. *They are commutative:*

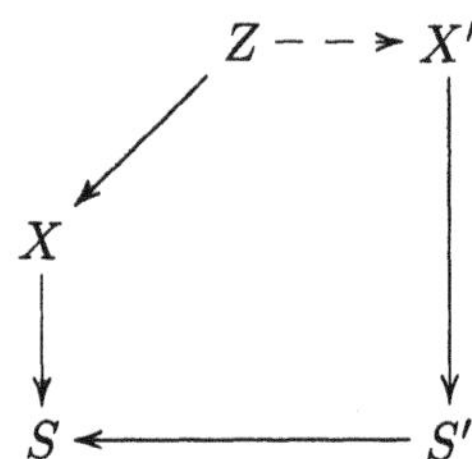

where $Z \to X$ is a Mori extremal divisorial contraction and $Z \dashrightarrow X'$ a sequence of Mori flips, flops or inverse Mori flips. Note that $\rho(S') - \rho(S) = 1$.

(4.2.2) Links of type II. *They are commutative:*

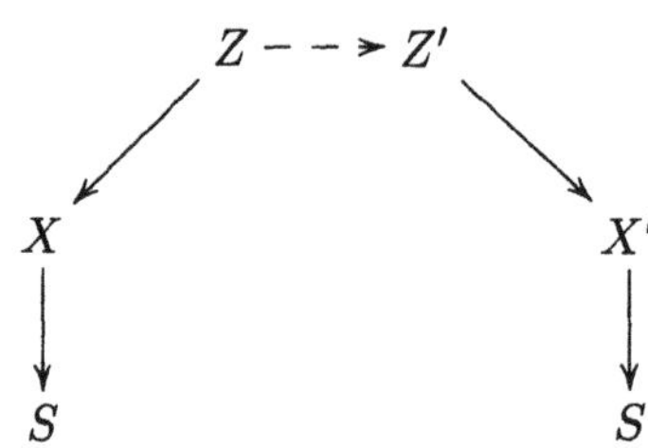

where $Z \to X$ and $Z' \to X'$ are Mori extremal divisorial contractions, and $Z \dashrightarrow Z'$ a sequence of Mori flips, flops or inverse Mori flips. The link induces here an isomorphism $S \cong S'$.

(4.2.3) Links of type III. *They are commutative:*

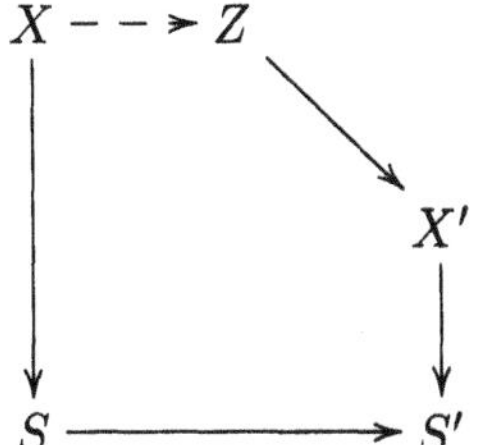

where $X \dashrightarrow Z$ is a sequence of Mori flips, flops or inverse Mori flips, and $Z \to X'$ a Mori extremal divisorial contraction. Note here that $\rho(S) - \rho(S') = 1$.

(4.2.4) Links of type IV. *They are commutative:*

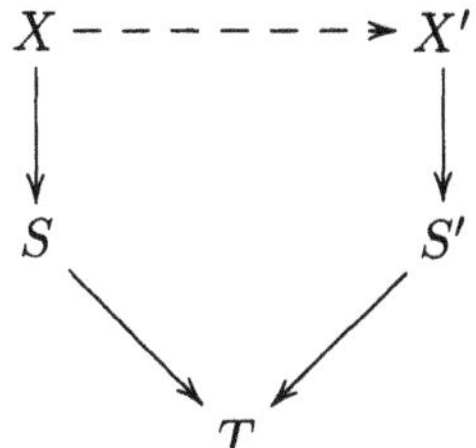

where $X \dashrightarrow X'$ is a sequence of Mori flips, flops or inverse Mori flips. Here $\rho(S) - \rho(T) = \rho(S') - \rho(T) = 1$.

PROOF. There are two cases:

a) If $K_X + \frac{1}{\mu}\mathcal{H}$ does not have canonical singularities, let $c < \frac{1}{\mu}$ be the maximum such that $K_X + c\mathcal{H}$ has canonical singularities. There is an extremal blow-up $f : Z \to X$ with exceptional E such that

$$K_Z + cf_*^{-1}\mathcal{H} = f^*(K_X + c\mathcal{H}).$$

A minimal model program for $K_Z + cf_*^{-1}\mathcal{H}$ over S produces the desired Φ_1 of type I or II.

b) If $K_X + \frac{1}{\mu}\mathcal{H}$ does have canonical singularities, either we are done (4.1.3) or $K_X + \frac{1}{\mu}\mathcal{H}$ is not nef. A minimal model program for $K_X + \frac{1}{\mu}\mathcal{H}$ gives Φ_1 of type III or IV. Care has to be paid not to lose S.

It remains to show that repeated application of steps (a) and (b) above eventually gives a factorization. It is easy to see that μ decreasesafter "untwisting" by Φ_1, and that it strictly decreases unless the base locus of $\mathcal{H}$ has improved by a measurable bit. These bits however may get smaller and smaller. In [C] this is the major technical difficulty. The proof uses results of [Ko, Ch. 18] and Alexeev, showing that certain sets satisfy the ascending chain condition. [Ko, Ch 18] contains some conjectures along these lines, [Ko4] explores a somewhat different direction (see also [A]). □

I will make some comments on the last statement and the future ambitions of this theory.

First, one should ask why the transformations I–IV above should be considered "elementary". Let me once again emphasize that the main philosophical point here is that the Mori fibration morphism is the relevant structure. So whatever an elementary map is, it has to "link" a Mori fibration $X \to S$ to another Mori fibration $X' \to S'$. So, for example, a blow-up $Z \to X$ of the maximal ideal of a smooth point in X should not be considered an elementary map, unless we are in the unlikely case that Z admits a Mori fibration $Z \to S'$ of its own. Elementary transformations of ruled surfaces then show that we need to allow at least two divisorial contractions to form an elementary map. Examples show that in dimension ≥ 3 flops and flips (hence inverse flips) are necessary. This explains that elementary maps satisfy some reasonable economy criterion.

It is to be hoped that the factorization theorem will eventually lead to usable criteria to determine the birational type of a given space admitting a Mori fibration. This definitely requires a better understanding of elementary maps. The idea is that the graph $\Gamma_\Phi \subset X \times X'$ of an elementary map $\Phi : X/S \dashrightarrow X'/S'$ is a relative $\mathbb{Q}$-Fano model with $\rho = 2$ (over S and/or S') and these tend to fit in a finite number of algebraic families. This however is not true because of the possibility of inverse flips, whose role has yet to be clarified. It seems that a close understanding (perhaps a classification) of divisorial contractions would also be necessary.

The following is taken from [I]:

(4.3) CONJECTURE. *Let $X \to S$ be a standard conic bundle, i.e., X is smooth and $X \to S$ the contraction of an extremal ray. Assume that X is rational. Then the quasieffective threshold $\tau = \tau(S, \Delta)$—that is, the maximum value τ such that $\tau K + \Delta$ is quasieffective—is less than* 2.

The idea is that the quasieffective threshold of the pair (S, Δ) has a strong birational meaning for X.

5. Log abundance

This is the statement of the log abundance theorem for threefolds, proved in [KMM].

(5.1) THEOREM. *Let (X, B) be a pair consisting of a threefold X and boundary*

$B \subset X$. If $K+B$ is log canonical and nef, $|m(K+B)|$ is free from base points for some sufficiently large (and divisible) m. □

The genuine (i.e., non logarithmic) abundance theorem states that if X is a three-dimensional minimal model (recall that this means that X has terminal singularities and K_X is nef), $|mK_X|$ is free from base points for some m large and divisible.

Recall that the *Kodaira dimension* $\kappa(D)$ of a divisor D on a variety X is the dimension of $\varphi_{mD}(X)$ for all sufficiently large and divisible integers m ($\varphi_{mD}: X \dashrightarrow \mathbb{P}H^0(mD)^*$ is the canonical map; by convention $\kappa(D) = -\infty$ if $H^0(mD) = (0)$ for all positive integers m). Also, the *numerical dimension* of a nef $\mathbb{Q}$-Cartier divisor D on a variety X is by definition the largest integer $\nu(D)$ such that $D^\nu \not\equiv 0$. It is easy to show that $\kappa \leq \nu$. It is also easy to see that log abundance is equivalent to $\nu(K+B) = \kappa(K+B)$. Log abundance for $\nu(K+B) = \dim(X)$ is an immediate consequence of the base point free theorem.

The proof of the (genuine) abundance theorem for threefolds (mainly due to Kawamata and Miyaoka) is very long and complicated: it can be found in [Ko], together with proper attributions. Roughly speaking, the proof is divided in two parts, requiring entirely different techniques. Here is a quick summary:

A) First one shows that $|mK| \neq \varnothing$ for some m. This is quite hard, and I refer to [Ko, Ch. 9] for an extremely attractive presentation.

B) We must eventually show that mK_X is free for m large and divisible. By (A) there is a divisor $D \in |mK|$. Note that if $\nu = 0$ we are done already, so we need to discuss $\nu = 1$ and $\nu = 2$. Roughly speaking, the argument proceeds by induction on the dimension: we show that a form of abundance holds on D, then try to use exact sequences to lift it to X. However, D may be too singular to work with. The first step is to modify X to improve the singularities of the support of D, namely to achieve that $K + D_{\text{red}}$ is log canonical. This is achieved by choosing a log resolution X', D' of X, D_{red}, and running a $(K_{X'}+D')$-minimal model program, ending in the new X'', D''. Then $K_{X''} + D''$ is log terminal and D'' is semi log canonical. So, perhaps at the expense of introducing slightly worse singularities on X, we may assume that $K_{D_{\text{red}}}$ is semi log canonical.

If $\nu = 1$ (here we expect the canonical map to give a fibration in surfaces), by a further reduction step, using again log minimal model theory, we may assume that every connected component of D_{red} is irreducible. Then D_{red} is a surface with log terminal singularities and it is not too hard to show that log abundance for D_{red} implies abundance for X. An excellent introduction to the yoga is [Ko, Ch. 11], where log abundance is proved for surfaces with log canonical singularities.

The situation is technically more complicated if $\nu = 2$. Here D_{red} is a surface with semilog canonical singularities. Abundance for D_{red} requires some careful considerations [Ko, Ch. 12]. To show abundance for X, it is enough to show that $H^0(mK_X)$ has at least two sections for m large (i.e., $\kappa \geq 1$). This uses log abundance on D_{red} and some delicate estimates of the relevant terms in the Riemann–Roch formula.

In closing, I wish to emphasize that even to prove genuine abundance one naturally dips in the log category.

The proof of the log abundance theorem for threefolds, part (B), follows closely the proof of genuine abundance, except that it is more difficult at various technical points, especially in the proof that $\nu = 2$ implies $\kappa \geq 1$ in showing the positivity of the relevant terms in the Riemann–Roch formula.

However, showing that $|m(K+B)| \neq \varnothing$ for some m requires some entirely new ideas, which I shall now describe following [KMM].

(5.2) THEOREM. *Let (X, B) be a pair consisting of a threefold X and boundary $B \subset X$. If $K+B$ is log canonical and nef, $|m(K+B)| \neq \varnothing$ for some positive integer m.*

PROOF. a) First construct a terminal modification $f : Z \to X$. This means that Z has terminal singularities,

$$K_Z + f_*^{-1}B + \sum a_i E_i = f^*(K_X + B)$$

with $0 \leq a_i \leq 1$ and $K_Z + B_0$ is log canonical when $B_0 = f_*^{-1}(B) + \sum a_i E_i$.

b) Next run an (ordinary) minimal model program for K_Z. If K_Z is not nef there is a smallest value $0 < \varepsilon \leq 1$ such that $K_Z + \varepsilon B_0$ is nef, and an extremal ray R with $K_Z \cdot R < 0$ and $(K_Z + \varepsilon B_0) \cdot R = 0$. Let $Z \dashrightarrow Z_1$ be the divisorial contraction or flip of R, and $B_1 \subset Z_1$ the proper transform. Then inductively define a chain

$$\cdots (Z_i, \varepsilon_i B_i) \dashrightarrow (Z_{i+1}, \varepsilon_{i+1} B_{i+1}) \dashrightarrow \cdots \dashrightarrow (Z_N, \varepsilon_N B_N) = (X', B'),$$

where $\varepsilon_i \leq \varepsilon_{i-1}$ is the smallest value such that $K_{Z_i} + \varepsilon_i B_i$ is nef, and $Z_i \dashrightarrow Z_{i+1}$ the divisorial contraction or flip associated to an extremal ray $R_i \subset \overline{NE}(Z_i)$ with $K_{Z_i} \cdot R_i < 0$ and $(K_{Z_i} + \varepsilon_i B_i) \cdot R_i = 0$. It is quite clear that $|m(K_Z + B_0)| \neq \varnothing$ for some m if $|m(K_{X'} + B')| \neq \varnothing$ for some m. There are two cases:

b1) X' is a minimal model and $B' = \varnothing$. Here $|mK_{X'}| \neq \varnothing$ by genuine abundance.

b2) X' has a Mori fibration $X' \to S$ with $(K_{X'} + B') \cdot C = 0$ for a curve C contained in a fiber. The result is immediate if X' is a $\mathbb{Q}$-Fano threefold, but the other cases still require considerable work. From now on, I assume (b2) holds.

c) The strategy at this point is to produce sections of $|m(K_{X'/S}+B')|$. It is easier to do this because the relative dualizing sheaf $K_{X'/S}$ behaves well with respect to fiber squares. That is, if $Y' = X' \times_S T$ in

$$\begin{array}{ccc} Y' & \xrightarrow{f'} & X' \\ \downarrow{\scriptstyle \varphi'} & & \downarrow{\scriptstyle \varphi} \\ T & \xrightarrow{f} & S \end{array}$$

then $K_{Y'/T} = f'^*K_{X'/S}$. By choosing a suitable finite $T \to S$ I can "untwist" X'/S, and going again through (b), I may assume that $X' \to S$ is a $\mathbb{P}^1$-bundle over a surface or a $\mathbb{P}^2$-bundle over a curve, and argue there directly.

d) Now assume $|m(K_{X'/S}+B')| \neq \varnothing$. This means that there is an effective $\mathbb{Q}$-divisor D on S such that $\varphi^*(D) = K_{X'/S}+B'$. Since $K_{X'} = K_{X'/S} + \varphi^*K_S$, this says that $\varphi^*(K_S+D) = K_{X'}+B'$. It can be shown that K_S+D is log terminal, and since it is nef, we are done by log abundance on S. □

The log abundance theorem is a very high-brow generalization of the genuine abundance theorem. Its meaning has to be found in the context of log minimal model theory, as a natural extension of minimal model theory. One very interesting feature peculiar to log abundance as opposed to genuine abundance is the analysis of Mori fibrations. The proof just described provides some new tools to study numerical aspects of curves and divisors on Mori fibrations. Let $X \to S$ be a Mori fibration, $\mathcal{H}$ a linear system without base divisors on X, and assume $K_X + \frac{1}{\mu}\mathcal{H}$ to be trivial on fibers. Theorem (5.2) suggests that $K_X + \frac{1}{\mu}\mathcal{H}$ is likely to be quasieffective, which explains the experimental fact that the possibility of X admitting a birationally distinct Mori fibration should be deemed unlikely (compare 4.1.2).

6. Effective base point freeness

A lot of work has been done recently on finding effective nonvanishing and global generation results for pluricanonical (nK) and adjoint ($K+L$) divisors in higher dimension [EL], [Ko2], and [Ko3]. The common ground of most of this work is an attempt to render the "Kawamata technique" effective.

This is a technically simplified version of the main result in [EL]:

(6.1) THEOREM. *Let L be a nef and big divisor on a smooth complex projective threefold X, and let $x \in X$ be a given point. Assume:*

(6.1.1) $L^3 > 92$,

(6.1.2) $L^2 \cdot S \geq 7$ *for all surfaces $S \subset X$ with $S \ni x$,*

(6.1.3) $L \cdot C \geq 3$ *for all curves $C \subset X$ with $C \ni x$.*

Then K_X+L is free at x, that is, $\mathcal{O}(K+L)$ has a section that is nonvanishing at x. Moreover, if $L^3 \gg 0$ (e.g., $L^3 \geq 1000$), the same conclusion holds with $L^2 \cdot S \geq 5$. □

The proof of the above result is quite complicated. The Kawamata technique is still in a stage where it is rather difficult to use it to prove things, even when one has a good idea of what should be proved. In the case at hand, the general form of the statement is suggested by earlier results of Reider on algebraic surfaces (proved with Bogomolov's stability of vector bundles). As an introduction to the ideas, following [EL], I will state and completely prove a Reider type statement on surfaces:

(6.2) THEOREM. *Let L be a nef and big divisor on a smooth complex projective surface X, and let $x \in X$ be a given point. Assume:*

(6.2.1) $L^2 \geq 5$,

(6.2.2) $L \cdot C \geq 2$ *for all curves $C \subset X$ with $C \ni x$.*

Then K_X+L is free at x, that is, $\mathcal{O}(K+L)$ has a section that is nonvanishing at x.

PROOF. One starts by writing the divisor we are interested in, $K_X + L$, in the form:

$$K + L = K + B + M$$

where $B \geq 0$ is a nef $\mathbb{Q}$-divisor, M a nef and big $\mathbb{Q}$-divisor. The method works by choosing B very singular at x, and relating the singularity of B at x to sections of $K + L$ at x.

The leading term of Riemann–Roch and (6.2.1) give $D \in |nL|$ with a singular point of multiplicity at least $2n + 1$ at x for some n large. Put $B = \frac{1}{n}D$, and write $B = B' + \sum b_i B_i$, where the B_i's are the irreducible components of B containing x. Let μ_i be the multiplicity of B_i at x. By construction,

$$b = \sum b_i \mu_i > 2 \qquad (*).$$

I discuss two cases:

a) $b > 2b_i$ for all i.

b) There is a value i_0, say $i_0 = 0$, such that $b \leq 2b_0$. In particular, $b_0 > 1$. What happens here is that we chose D to be very singular at x, and unexpectedly got a D that has very high multiplicity along all of B_0. It seems that this occurrence should be considered lucky, but it is the hardest to work with.

a) Let $f : Z \to X$ be the blow-up at x, and E the exceptional curve. Then:

$$f^* K_X = K_Z - E$$

and $f^*B = f_*^{-1}B + bE$, so for all c we have

$$\begin{aligned} f^*(K_X + L) &= f^*(K_X + cB) + (1-c)f^*B \\ &= K_Z + cf_*^{-1}B + (cb-1)E + (1-c)f^*B. \end{aligned}$$

Set $c = 2/b$. Then $M = (1-c)f^*B$ is a nef and big $\mathbb{Q}$-divisor. Also, by assumption, $cb_i < 1$. Then

$$N = f^*(K+L) - \lfloor cf_*^{-1}B \rfloor = K_Z + \lceil M \rceil + E.$$

By Kodaira vanishing $H^0(N) \twoheadrightarrow H^0(N|_E)$. We are done because $N|_E \sim \mathcal{O}_E$ and $H^0(K_X + L) \hookrightarrow H^0(N|_E)$.

b) In this case with $c = 1/b_0$ and $M = (1-c)B$,

$$N = K_X + L - \lfloor cB' \rfloor = K_X + \lceil M \rceil + B_0.$$

By Kodaira vanishing $H^0(N) \twoheadrightarrow H^0(N|_{B_0})$, so we are done as soon as we prove that $H^0(N|_{B_0})$ has a section not vanishing at x. This is the ugly part, where it is very helpful to already know the statement we wish to prove. The observation here is that

$$N|_{B_0} = K_{B_0} + (1-c)B|_{B_0} + \sum_{i>0} cb_iB_i|_{B_0}$$

is an integral divisor, and

$$[(1-c)B + \sum_{i>0} cb_iB_i] \cdot B_0 > 2(1-c) + \sum_{i>0} cb_i\mu_i = 1 + c(\sum_{i\geq 0} b_i\mu_i - 2) > 1$$

by (*) applied twice. The statement then follows from the one-dimensional analogue of (6.2) (more precisely, one needs (6.2) for Gorenstein curves). □

In higher dimensions matters quickly become more complicated, seemingly due to the possibility of case (b) above, especially in conjunction to the more complicated, but necessary in dimension ≥ 3, hypotheses in the Kawamata–Viehweg vanishing theorem. The reader may show as an exercise that case (b) never occurs if $L \cdot C \geq 2$ is replaced with $L \cdot C \geq 3$ in (6.2.2) (I learned this observation from R. Lazarsfeld, who attributes it to Demailly).

Finally, I wish to make a philosophical remark. The log category intervened "behind the scenes" in the proof of (6.2). Indeed, we chose B such that $K + B$ is not log canonical at x, and c the maximum such that $K + cB$ is log canonical at x, in order to isolate the base component with "maximal multiplicity".

References

[A] V. Alexeev, *Boundedness and K^2 for log surfaces*, preprint (1994).

[C] A. Corti, *Factoring birational maps of threefolds after Sarkisov*, J. Algebraic Geom. (to appear).

[EL] L. Ein and R. Lazarsfeld, *Global generation of pluricanonical and adjoint linear series on smooth projective threefolds*, J. Amer. Math. Soc. **6** (1993), 875–903.

[I] V. A. Iskovskikh, *Towards the problem of rationality of conic bundles*, Algebraic Geometry, Proceedings of the US-USSR Symposium, Chicago 1989, Lecture Notes in Mathematics **1479**, Springer-Verlag, New York, pp. 50–56.

[Ka] Y. Kawamata, *Semistable minimal models of threefolds in positive or mixed characteristic*, J. Algebraic Geom. **3** (1994), 463–491.

[KMM] S. Keel, J. McKernan and K. Matsuki, *Log abundance theorem for threefolds*, Duke Math. J. **75** (1994), 99–119.

[Ko] J. Kollár et al., *Flips and abundance for algebraic threefolds*, Astérisque **211** (1992).

[Ko2] J. Kollár, *Effective base point freeness*, Math. Ann. **296** (1993), 595–605.

[Ko3] ———, *Shafarevich maps and plurigenera of algebraic varieties*, Invent. Math. **113** (1993), 177–215.

[Ko4] ———, *Log surfaces of general type: some conjectures*, in *Classification of algebraic varieties*, Contemp. Math. **162** (1994), Amer. Math. Soc., 261–276.

[S] V. V. Shokurov, *Semi-stable 3-fold flips*, Russian Acad. Sci. Izv. Math. **42** (1994), 371–425.

Alessio Corti
Department of Mathematics, University of Chicago,
5734 S. University Avenue
Chicago IL 60637

E-mail address: corti@math.uchicago.edu

Complex Algebraic Geometry
MSRI Publications
Volume **28**, 1995

The Schottky Problem: An Update

OLIVIER DEBARRE

ABSTRACT. The aim of these lecture notes is to update Beauville's beautiful 1987 Séminaire Bourbaki talk on the same subject. The Schottky problem is the problem of finding characterizations of Jacobians among all principally polarized abelian varieties. We review the numerous approaches to this problem. In the "analytical approach", one tries to find polynomials in the thetaconstants that define the Jacobian locus in the moduli space of principally polarized abelian varieties. We review van Geemen's (1984) and Donagi's (1987) work in that direction. The loci they get contain the Jacobian locus as an irreducible component. In the "geometrical approach", one tries to give geometric properties that are satisfied only by Jacobians. We review the following: singularities of the theta divisor (Andreotti and Mayer 1967); reducibility of intersections of a theta divisor with a translate (Welters 1984) and trisecants to the Kummer variety (Welters 1983, Beauville and Debarre 1986, Debarre 1992); the K–P equation and Novikov's conjecture (Shiota 1986, Arbarello and De Concini 1984); double translation hypersurfaces (Little 1989); the van Geemen–van der Geer conjectures on the base locus of the set of second order theta functions that vanish with multiplicity ≥ 4 at the origin (Beauville and Debarre 1989, Izadi 1993); subvarieties with minimal class (Ran 1981, Debarre 1992); the Buser–Sarnak approach (1993).

Introduction

For $g \geq 2$, the moduli space $\mathcal{A}_g$ of principally polarized abelian varieties of dimension g has dimension $\frac{1}{2}g(g+1)$ and the moduli space $\mathcal{M}_g$ of smooth connected projective algebraic curves of genus g has dimension $3g-3$. To any such curve C is associated a principally polarized abelian variety of dimension g, its Jacobian (JC, θ). This defines a map

$$\mathcal{M}_g \longrightarrow \mathcal{A}_g,$$

Partially supported by NSF Grant DMS 9203919 and the European Science Project "Geometry of Algebraic Varieties", Contract no. SCI-0398-C (A).

which is injective (Torelli theorem). The closure $\mathcal{J}_g$ of its image is equal to $\mathcal{A}_g$ only for $g = 2$ or 3. For $g \geq 4$, it is a proper closed subset. The Schottky problem is the problem of finding characterizations of Jacobians among all principally polarized abelian varieties (for $g \geq 4$). For the sake of simplicity, we will say that a property is a *weak characterization* of Jacobians if $\mathcal{J}_g$ is an irreducible component of the set of principally polarized abelian varieties with this property.

1. The analytical approach

Certain modular forms, called "thetaconstants", define an embedding of a finite cover of $\mathcal{A}_g$ into some projective space [I]. The idea is to use the corresponding "coordinates" to give equations of $\mathcal{J}_g$ in $\mathcal{A}_g$. A brief history of this approach goes as follows:

Schottky [S, 1888] found a polynomial of degree 16 in the thetaconstants that vanishes on $\mathcal{J}_4$ but not on $\mathcal{A}_4$.

Schottky and Jung [SJ, 1909], starting from polynomials in the thetaconstants that vanish on $\mathcal{A}_{g-1}$, constructed polynomials that vanish on $\mathcal{J}_g$.

Igusa [I, 1981] and **Freitag** [F, 1983] proved that the divisor defined by the Schottky polynomial is *irreducible*, hence equal to $\mathcal{J}_4$.

Van Geemen [vG, 1984] proved that $\mathcal{J}_g$ is an irreducible component of the locus $\mathcal{S}_g$ defined by the Schottky–Jung polynomials.

Donagi [Do1, 1984] remarked that the Schottky–Jung polynomials depend on the choice of a point of order two on the abelian variety, hence that they define in fact two loci $\mathcal{S}_g$ and $\mathcal{S}_g^{\text{big}}$, depending on whether one considers all or just one single such point. They satisfy

$$\mathcal{J}_g \subset \mathcal{S}_g \subset \mathcal{S}_g^{\text{big}}.$$

While van Geemen proved that $\mathcal{J}_g$ is a component of $\mathcal{S}_g$, Donagi proved that it is also a component of $\mathcal{S}_g^{\text{big}}$, and that $\mathcal{J}_g \neq \mathcal{S}_g^{\text{big}}$ for $g \geq 5$ [Do2]. For example, intermediate Jacobians of cubic threefolds are in $\mathcal{S}_5^{\text{big}}$, but not in $\mathcal{S}_5$. Donagi has a conjecture on the structure of $\mathcal{S}_g^{\text{big}}$ for any g, which would imply $\mathcal{S}_g = \mathcal{J}_g$. He has announced in [Do3] a proof of this conjecture for $g = 5$.

2. The geometrical approach

The idea here is to give geometric properties of a principally polarized abelian variety that are satisfied only by Jacobians.

a) Singularities of the theta divisor. If Θ is a theta divisor on a g-dimensional Jacobian, $\dim \operatorname{Sing} \Theta \geq g - 4$. Andreotti and Mayer proved [AM, 1967] that this property is a *weak characterization* of Jacobians. However, the locus in $\mathcal{A}_g$ defined by this property always has components other than $\mathcal{J}_g$ for $g \geq 4$, although $\mathcal{J}_g$ is the only known component that is not contained in θ_{null} (the divisor in $\mathcal{A}_g$ of principally polarized abelian varieties for which a theta-constant vanishes) [D1].

b) Reducibility of $\Theta \cap \Theta_a$ and trisecants. Weil observed in [W] that if Θ is a theta divisor on the Jacobian JC of a smooth curve C, then, for any points p, q, r and s of C, one has the inclusion

$$\Theta \cap \Theta_{p-q} \subset \Theta_{p-r} \cup \Theta_{s-q}, \qquad (*)$$

where Θ_x stands for the translate $\Theta + x$. Now let (A, θ) be an indecomposable principally polarized abelian variety and let Θ be a symmetric theta divisor. Assume that

$$\Theta \cap \Theta_a \subset \Theta_x \cup \Theta_y$$

for some distinct non-zero points a, x and y of A (in particular, $\Theta \cap \Theta_a$ is reducible). This inclusion has a nice geometric interpretation in terms of the *Kummer map* $K : A \to |2\Theta|^*$ associated with the linear system $|2\Theta|$: for any point ζ of A such that $2\zeta = x + y$, it is equivalent to the fact that $K(\zeta)$, $K(\zeta - a)$ and $K(\zeta - x)$ are *collinear*. In other words, the Kummer variety $K(JC)$ of a Jacobian has a family of dimension 4 of *trisecant lines*. Mumford suggested that this property should characterize Jacobians. The most ambitious version of this conjecture is due to Welters [We1]:

CONJECTURE. *If the Kummer variety of an indecomposable principally polarized abelian variety (A, θ) has one trisecant line, then (A, θ) is a Jacobian.*

The following is known:

Welters [We2, 1983], using Gunning's criterion [Gu], showed that if $K(A)$ has a one-dimensional family of trisecants and $\dim \operatorname{Sing} \Theta \leq \dim A - 4$, then (A, θ) is a Jacobian.

Beauville and Debarre [BD1, 1986] showed that if $K(A)$ has one trisecant, then $\dim \operatorname{Sing} \Theta \geq \dim A - 4$. This result, combined with the result of Andreotti–Mayer mentioned above, implies that the existence of one trisecant is a weak characterization of Jacobians.

Debarre [D2, 1989] proved that Welters' conjecture holds for $g \leq 5$. Moreover, if (A, θ) is indecomposable, if $K(a)$, $K(b)$ and $K(c)$ are collinear and if the subgroup of A generated by $a - b$ and $a - c$ is dense in A, then (A, θ) is a Jacobian [D3, D4].

c) The K–P equation. If one lets p, q, r and s go to the same point of C in $(*)$, one gets that a theta function of a Jacobian satisfies the K–P equation, a non-linear partial differential equation that depends on three constant vector fields.

Shiota [Sh, 1986] proved that the K–P equation characterizes Jacobians among all indecomposable principally polarized abelian varieties.

Shiota's proof is analytical. It was later partially algebraized and simplified by Arbarello and De Concini [AD], but they could not bypass a crucial point in Shiota's proof. At this point, all existing algebraic proofs need extra hypotheses.

d) Double translation type. If Θ is a symmetric theta divisor on the Jacobian of a curve C of genus g, the existence of an Abel–Jacobi map $C^{g-1} \to \Theta$ implies that Θ can be locally parametrized by

$$(t_1, \dots, t_{g-1}) \mapsto \alpha_1(t_1) + \dots + \alpha_{g-1}(t_{g-1}),$$

where $\alpha_i : \mathbf{C} \to \mathbf{C}^g$ are curves. By symmetry of Θ, there are actually two such parametrizations if C is not hyperelliptic, and the theta divisor of a Jacobian is said to be a *double translation hypersurface.* The following is known:

Lie [L1, L2, 1935] and **Wirtinger** [Wi] proved that a non-developable (having a generically finite Gauss map) double translation hypersurface in $\mathbf{C}^g$ is a piece of the theta divisor of a Gorenstein curve of arithmetic genus g.

Little [Li1, 1989] later showed that if the theta divisor of a principally polarized abelian variety is locally a *generalized* double translation hypersurface, that is, has two local parametrizations of the type

$$(t_1, \dots, t_{g-1}) \mapsto \alpha(t_1) + A(t_2, \dots, t_{g-1}),$$

the principally polarized abelian variety is a Jacobian. He recently made the connection with other approaches by showing that if the Kummer variety of an indecomposable complex principally polarized abelian variety has a "curve of flexes" (that is, satisfies a one-dimensional family of K–P equations), the theta divisor is a generalized translation hypersurface [Li2].

e) Second order theta functions. Let (A, θ) be a principally polarized abelian variety of dimension g. The linear system $|2\Theta|$ is independent of the choice of a *symmetric* theta divisor Θ. It has dimension $2^g - 1$. When (A, θ) is indecomposable, the subsystem $|2\Theta|_{00}$ of divisors that have multiplicity ≥ 4 at 0 has codimension $\frac{1}{2}g(g+1)$. When (A, θ) is the Jacobian of a smooth curve C, Welters showed [We3] that the base locus of $|2\Theta|_{00}$ is (as a set) the surface $C - C$ in JC (except possibly when $g = 4$, when there might be two extra isolated points). One might hope to use this property to characterize Jacobians:

CONJECTURE [vGvG]. *If the indecomposable principally polarized abelian variety (A, θ) is not a Jacobian, the base locus of $|2\Theta|_{00}$ is $\{0\}$ as a set.*

As explained in [BD2], this conjecture is connected to the trisecant conjecture in the following way: the linear system $|2\Theta|_{00}$ corresponds to hyperplanes in $|2\Theta|^*$ that contain the point $K(0)$ and the Zariski tangent space $T_{K(0)}K(A)$. Its base locus is therefore the inverse image by the Kummer map of $K(A) \cap T_{K(0)}K(A)$. The conjecture says that if (A, θ) is not a Jacobian, there are no lines through $K(0)$ and another point of $K(A)$ that are contained in $T_{K(0)}K(A)$. Note that such a line meets $K(A)$ with multiplicity ≥ 3. What is known is the following:

Beauville and Debarre [BD2, 1989] proved that if (A, θ) is generic of dimension ≥ 4, then the base locus of $|2\Theta|_{00}$ is finite. Also, the conjecture holds for the intermediate Jacobian of a cubic threefold and various other examples.

Izadi's [Iz, 1993] work on $\mathcal{A}_4$ implies the conjecture for $g = 4$.

This conjecture has an infinitesimal analog. By associating to an element of $|2\Theta|_{00}$ the fourth-order term of the Taylor expansion of a local equation at 0, we get a linear system of quartics in $\mathbf{P}T_0A$. Using a result of Green [G], Beauville showed that, for the Jacobian of a smooth curve C, the base locus of this linear system is equal to the canonical curve of C (except possibly when $g = 4$, when there might be an extra isolated point) [BD2]. Again, one might hope to use this property to characterize Jacobians:

CONJECTURE [vGvG]. *If the indecomposable principally polarized abelian variety (A, θ) is not a Jacobian, this linear system of quartics is base-point-free.*

Donagi explains in [Do3] the relationship between this conjecture and the K–P equation: if D_1 is a base point, the second-order theta functions θ_n satisfy an equation of the type

$$(D_1^4 + \text{ lower order terms })\,\theta_n(z, \tau) = 0,$$

whereas the K–P equation is equivalent to

$$(D_1^4 - D_1D_3 + D_2^2 + d)\,\theta_n(z, \tau) = 0.$$

The following is known:

Beauville–Debarre [BD2, 1989] proved the conjecture for a generic principally polarized abelian variety of dimension ≥ 4, for the intermediate Jacobian of a cubic threefold and for various other examples.

Izadi's [Iz, 1993] work on $\mathcal{A}_4$ implies the conjecture for $g = 4$.

The difficulty with these conjectures is that the only known efficient way to construct elements of $|2\Theta|_{00}$ is to use singular points of the theta divisor (if a is singular on Θ, then $\Theta_a + \Theta_{-a}$ is in $|2\Theta|_{00}$). But in general Θ is smooth!

f) Subvarieties with minimal classes. If (JC, θ) is the Jacobian of a smooth curve C of genus g, the curve C embeds (non-canonically) into JC by the Abel–Jacobi map. The image has cohomology class θ_{g-1} (for any integer d, we write θ_d for the *minimal* (that is, non-divisible) cohomology class $\theta^d/d!$). The existence of such a curve characterizes Jacobians (Matsusaka's criterion, [M]). This result has been improved on:

Ran [R, 1981] (see also [C]), proved that if an irreducible curve C generates a g-dimensional principally polarized abelian variety (A, θ) and satisfies $C \cdot \theta = g$, then C is smooth and (A, θ) is isomorphic to its Jacobian.

More generally, for any $d \leq g$, the symmetric product $C^{(d)}$ maps onto a subvariety $W_d(C)$ of JC with cohomology class θ_{g-d}. However, the existence of a subvariety with minimal cohomology class does *not* characterize Jacobians: the Fano surface of lines on a cubic threefold maps onto a surface with class θ_3 in the intermediate Jacobian [CG, B2]. Nonetheless, the following is known:

Ran [R, 1981] proved that if a principally polarized abelian fourfold contains a surface with class θ_2, it is the Jacobian of a curve C of genus 4 and the surface is a translate of $\pm W_2(C)$.

Debarre [D5, 1992] proved that the existence of a subvariety of codimension ≥ 2 with minimal class is a weak characterization of Jacobians. Moreover, the only subvarieties of a Jacobian JC with minimal classes are translates of $\pm W_d(C)$.

A natural extension of Ran's result would be:

CONJECTURE. *If the g-dimensional principally polarized abelian variety (A, θ) contains subvarieties V and W with respective cohomology classes θ_d and θ_{g-d}, then (A, θ) is a Jacobian.*

One can be even more ambitious. Let V be a subvariety of dimension d of a principally polarized abelian variety (A, θ). Call V non-degenerate if the restriction map:

$$H^0(A, \Omega_A^d) \longrightarrow H^0(V_{\text{reg}}, \Omega_{V_{\text{reg}}}^d)$$

is injective. For example, a curve is non-degenerate if and only if it generates A. A divisor is non-degenerate if and only if it is ample. Moreover, any subvariety with class a multiple of a minimal class is non-degenerate. Ran's above characterization of Jacobian fourfolds actually holds under the weaker hypotheses that the surface is non-degenerate and of self-intersection 6. In general, since $\theta_{g-d} \cdot \theta_d = \binom{g}{d}$, a nice generalization of Ran's results would be:

CONJECTURE. *If the g-dimensional principally polarized abelian variety (A, θ) contains non-degenerate subvarieties V and W of respective dimensions d and $g-d$ such that $V \cdot W = \binom{g}{d}$, then (A, θ) is a Jacobian.*

g) Buser–Sarnak's approach. A complex abelian variety (A, θ) can be written as the quotient of its universal cover $V \simeq \mathbf{C}^g$ by a lattice L. A polarization induces a positive definite Hermitian form H on V, whose real part B is symmetric positive definite. Set

$$\delta(A) = \min_{x \in L, x \neq 0} B(x, x).$$

Buser and Sarnak show in [BS] that for any g–dimensional Jacobian JC, one has

$$\delta(JC) \leq \frac{3}{\pi} \log(4g + 3).$$

On the other hand,

$$\max_{A \in \mathcal{A}_g} \delta(A) \geq \left(\frac{\pi^g}{2g!} \right)^{-1/g} \simeq \frac{g}{\pi e}.$$

In other words, the maximum of δ on $\mathcal{A}_g$ is much larger than its maximum on $\mathcal{M}_g$. This leads to an effective criterion for determining if a given lattice is not a Jacobian.

References

[AM] A. Andreotti and A. Mayer, *On period relations for abelian integrals on algebraic curves*, Ann. Sc. Norm. Sup. Pisa **21** (1967), 189–238.

[AD] E. Arbarello and C. de Concini, *Another proof of a conjecture of S.P. Novikov on periods of abelian integrals on Riemann surfaces*, Duke Math. J. **54** (1984), 163–178.

[B1] A. Beauville, *Le problème de Schottky et la conjecture de Novikov*, Séminaire Bourbaki, Exposé 675, 1986–1987.

[B2] ———, *Sous-variétés spéciales des variétés de Prym*, Comp. Math. **45** (1982), 357–383.

[BD1] A. Beauville and O. Debarre, *Une relation entre deux approches du problème de Schottky*, Invent. Math. **86** (1986), 195–207.

[BD2] ———, *Sur les fonctions thêta du second ordre*, Arithmetic of Complex Manifolds, Proceedings, Erlangen 1988,, Lecture Notes in Mathematics **1399**, Springer, 1989.

[BS] P. Buser and P. Sarnak, *Period Matrices of Surfaces of Large Genus*, with an appendix by Conway, J. and Sloane, N., Invent. Math. **117** (1994), 27–56.

[CG] C. H. Clemens and P. A. Griffiths, *The Intermediate Jacobian of the Cubic Threefold*, Ann. of Math. **95** (1972), 281–356.

[C] A. Collino, *A new proof of the Ran–Matsusaka criterion for Jacobians*, Proc. Amer. Math. Soc. **92** (1984), 329–331.

[D1] O. Debarre, *Sur les variétés abéliennes dont le diviseur thêta est singulier en codimension 3*, Duke Math. J. **56** (1988), 221–273.

[D2] ———, *The trisecant conjecture for Pryms*, Theta Functions, Bowdoin 1987, Proceedings of Symposia in Pure Mathematics 49, Part 1, 1989.

[D3] ———, *Trisecant Lines And Jacobians*, J. Alg. Geom. **1** (1992), 5–14.

[D4] ———, *Trisecant Lines And Jacobians*, II (to appear in J. Alg. Geom. 1995).

[D5] ———, *Minimal cohomology classes and Jacobians*, preprint.

[Do1] R. Donagi, *Big Schottky*, Invent. Math. **89** (1988), 569–599.

[Do2] ———, *Non-Jacobians in the Schottky loci*, Ann. of Math. **126** (1987), 193–217.

[Do3] ———, *The Schottky problem*, Theory of moduli, Proceedings, Montecatini Terme 1985, Lecture Notes in Mathematics *1337*, Springer, 1988.

[F] E. Freitag, *Die Irreduzibilität der Schottkyrelation*, Arch. Math. **40** (1983), 255–259.

[G] M. Green, *Quadrics of rank four in the ideal of a canonical curve*, Invent. Math. **75** (1984), 85–104.

[Gu] R. Gunning, *Some curves in abelian varieties*, Invent. Math. **66** (1982), 377–389.

[I] J. Igusa, *On the irreducibility of the Schottky divisor*, J. Fac. Sci. Tokyo **28** (1981), 531–545.

[Iz] E. Izadi, *The geometric structure of $\mathcal{A}_4$, the structure of the Prym map, double solids and Γ_{00}-divisors*, preprint.

[L1] S. Lie, *Die Theorie der Translationsflachen und das Abelsche Theorem*, Gesammelte Abhandlungen, Band II, Abh. XIII, Teubner, Leipzig, 1935.

[L2] ———, *Das Abelsche Theorem und die Translationsmannigfaltigkeiten*, Gesammelte Abhandlungen, Band II, Abh. XIV, Teubner, Leipzig, 1935.

[Li1] J. Little, *Translation manifolds and the Schottky problem*, Theta Functions, Bowdoin 1987, Proceedings of Symposia in Pure Mathematics 49, Part 1, 1989.

[Li2] ———, *Another Relation Between Approaches to the Schottky problem*, preprint.

[M] T. Matsusaka, *On a characterization of a Jacobian Variety*, Mem. Coll. Sc. Kyoto, Ser. A **23** (1959), 1–19.

[R] Z. Ran, *On subvarieties of abelian varieties*, Invent. Math. **62** (1981), 459–479.

[S] F. Schottky, *Zur Theorie der Abelschen Funktionen von vier Variabeln*, J. reine angew. Math. **102** (1888), 304–352.

[SJ] F. Schottky and H. Jung, *Neue Sätze über Symmetralfunktionen und die Abel'schen Funktionen der Riemann'schen Theorie*, S.-B. Preuss. Akad. Wiss., Berlin; Phys. Math. Kl. 1 (1909), 282–297.

[Sh] T. Shiota, *Characterization of Jacobian varieties in terms of soliton equations*, Invent. Math. **83** (1986), 333–382.

[vG] B. van Geemen, *Siegel modular forms vanishing on the moduli space of abelian varieties*, Invent. Math. **78** (1984), 329–349.

[vGvG] B. van Geemen and G. van der Geer, *Kummer varieties and the moduli spaces of abelian varieties*, Amer. J. of Math. **108** (1986), 615–642.

[W] A. Weil, *Zum Beweis des Torellischen Satzes*, Nach. Akad. Wiss. Göttingen; Math. Phys. Kl. 2a (1957), 33–53.

[We1] G. Welters, *A criterion for Jacobi varieties*, Ann. Math. **120** (1984), 497–504.

[We2] ———, *A characterization of non-hyperelliptic Jacobi varieties*, Invent. Math. **74** (1983), 437–440.

[We3] ———, *The surface $C - C$ on Jacobi varieties and second order theta functions*, Acta Math. **157** (1986), 1–22.

[Wi] W. Wirtinger, *Lies Translationsmannigfaltigkeiten und Abelsche Integrale*, Monatsh. Math. Phys. **46** (1938), 384–431.

Olivier Debarre
IRMA–Université Louis Pasteur–CRNS
7, rue René Descartes
67084 Strasbourg Cédex – France

E-mail address: debarre@math.u-strasbg.fr

Complex Algebraic Geometry
MSRI Publications
Volume **28**, 1995

Spectral Covers

RON DONAGI

ABSTRACT. Spectral curves have acquired a central role in understanding the moduli spaces of vector bundles and Higgs bundles on a curve. A Higgs G-bundle on an arbitrary variety S (together with some additional data, such as a representation of G) determines a *spectral cover* $\tilde{S}$ of S and an equivariant sheaf on $\tilde{S}$. The purpose of these notes is to combine and review various results about spectral covers, focusing on the decomposition of their Picards (and the resulting Prym identities) and the interpretation of a distinguished Prym component as parameter space for Higgs bundles.

1. Introduction

Spectral curves arose historically out of the study of differential equations of Lax type. Following Hitchin's work [H1], they have acquired a central role in understanding the moduli spaces of vector bundles and Higgs bundles on a curve. Simpson's work [S] suggests a similar role for spectral covers $\widetilde{S}$ of higher-dimensional varieties S in moduli questions for bundles on S.

The purpose of these notes is to combine and review various results about spectral covers, focusing on the decomposition of their Picards (and the resulting Prym identities) and the interpretation of a distinguished Prym component as parameter space for Higgs bundles. Much of this is modeled on Hitchin's system, which we recall in Section 1, and on several other systems based on moduli of Higgs bundles, or vector bundles with twisted endomorphisms, on curves. By peeling off several layers of data that are not essential for our purpose, we arrive at the notions of an *abstract principal Higgs bundle* and a *cameral* (roughly, a principal spectral) cover. Following [D3], this leads to the statement of the main result, Theorem 12, as an equivalence between these somewhat abstract 'Higgs' and 'spectral' data, valid over an arbitrary complex variety and for a reductive Lie group G. Several more familiar forms of the equivalence can then be derived in special cases by adding choices of representation, value bundle

Partially supported by NSF Grant DMS 95–03249 and NSA GRant MDA 904–92–H3047.

and twisted endomorphism. This endomorphism is required to be *regular*, but not semesimple. Thus the theory works well even for Higgs bundles that are everywhere nilpotent. After touching briefly on the symplectic side of the story in Section 6, we discuss some of the issues involved in removing the regularity assumption, as well as some applications and open problems, in Section 7.

This survey is based on talks at the Vector Bundle Workshop at UCLA (October 92) and the Orsay meeting (July 92), and earlier talks at Penn, UCLA and MSRI. I would like to express my thanks to Rob Lazarsfeld and Arnaud Beauville for the invitations, and to them and Ching Li Chai, Phillip Griffiths, Nigel Hitchin, Vasil Kanev, Ludmil Katzarkov, Eyal Markman, Tony Pantev, Emma Previato and Ed Witten for stimulating and helpful conversations.

We work throughout over $\mathbf{C}$. The total space of a vector bundle (= locally free sheaf) K is denoted $\mathbb{K}$. Some more notation:

Groups:	$G\ B\ T\ N\ C$
algebras:	$\mathfrak{g}\ \mathfrak{b}\ \mathfrak{t}\ \mathfrak{n}\ \mathfrak{c}$
Principal bundles:	$\mathcal{G}\ \mathcal{B}\ \mathcal{T}\ \mathcal{N}\ \mathcal{C}$
bundles of algebras:	$\mathrm{g}\ \mathrm{b}\ \mathrm{t}\ \mathrm{n}\ \mathrm{c}$

2. Hitchin's system

Let $\mathcal{M} := \mathcal{M}_C(n,d)$ be the moduli space of stable vector bundles of rank n and degree d on a smooth projective complex curve C. It is smooth and quasiprojective of dimension

$$\tilde{g} := n^2(g-1)+1. \tag{1}$$

Its cotangent space at a point $E \in \mathcal{M}$ is given by

$$T^*_E\mathcal{M} := H^0(\mathrm{End}(E) \otimes \omega_C), \tag{2}$$

where ω_C is the canonical bundle of C. Our starting point is:

THEOREM 1 (HITCHIN [H1]). *The cotangent bundle $T^*\mathcal{M}$ is an algebraically completely integrable Hamiltonian system.*

Complete integrability means that there is a map

$$h : T^*\mathcal{M} \longrightarrow B$$

to a $\tilde{g}$-dimensional vector space B that is Lagrangian with respect to the natural symplectic structure on $T^*\mathcal{M}$ (i.e., the tangent spaces to a general fiber $h^{-1}(a)$ over $a \in B$ are maximal isotropic subspaces with respect to the symplectic form). In this situation one gets, by contraction with the symplectic form, a trivialization of the tangent bundle:

$$T_{h^{-1}(a)} \xrightarrow{\approx} \mathcal{O}_{h^{-1}(a)} \otimes T_a^* B. \tag{3}$$

In particular, this produces a family of ('*Hamiltonian*') vector fields on $h^{-1}(a)$ that is parametrized by T_a^*B, and the flows generated by these fields on $h^{-1}(a)$ all commute. *Algebraic complete integrability* means additionally that the fibers $h^{-1}(a)$ are Zariski open subsets of abelian varieties on which the Hamiltonian flows are linear, i.e., the vector fields are constant.

We describe the idea of the proof in a slightly more general setting, following [BNR]. Let K be a line bundle on C, with total space $\mathbb{K}$. (In Hitchin's situation, K is ω_C and $\mathbb{K}$ is T^*C.) A *K-valued Higgs bundle* is a pair

$$(E,\ \phi : E \longrightarrow E \otimes K)$$

consisting of a vector bundle E on C and a K-valued endomorphism. One imposes an appropriate stability condition, and obtains a good moduli space $\mathcal{M}_K$ parametrizing equivalence classes of K-valued semistable Higgs bundles, with an open subset $\mathcal{M}_K^s$ parametrizing isomorphism classes of stable ones [S].

Let $B := B_K$ be the vector space parametrizing polynomial maps

$$p_a : \mathbb{K} \longrightarrow \mathbb{K}^n$$

of the form

$$p_a(x) = x^n + a_1 x^{n-1} + \cdots + a_n, \qquad a_i \in H^0(K^{\otimes i}).$$

In other words,

$$B := \bigoplus_{i=1}^{n} H^0(K^{\otimes i}). \tag{4}$$

The assignment

$$(E, \phi) \longmapsto \operatorname{char}(\phi) := \det(xI - \phi) \tag{5}$$

gives a morphism

$$h_K : \mathcal{M}_K \longrightarrow B_K. \tag{6}$$

In Hitchin's case, the desired map h is the restriction of h_{ω_C} to $T^*\mathcal{M}$, which is an open subset of $\mathcal{M}_{\omega_C}^s$. Note that in this case $\dim B$ is, in Hitchin's words, 'somewhat miraculously' equal to $\tilde{g} = \dim \mathcal{M}$.

The *spectral curve* $\widetilde{C} := \widetilde{C}_a$ defined by $a \in B_K$ is the inverse image in $\mathbb{K}$ of the 0-section of $\mathbb{K}^{\otimes n}$ under $p_a : \mathbb{K} \longrightarrow \mathbb{K}^n$. It is finite over C of degree n. The general fiber of h_K is given by

PROPOSITION 2. [BNR] *For* $a \in B$ *with* integral *spectral curve* $\widetilde{C}_a$, *there is a natural equivalence between isomorphism classes of*

(i) *rank-1, torsion-free sheaves on* $\widetilde{C}_a$, *and*
(ii) *pairs* $(E\ ,\ \phi : E \to E \otimes K)$ *with* $\mathrm{char}(\phi) = a$.

When $\widetilde{C}_a$ is non-singular, the fiber is thus $\mathrm{Jac}(\widetilde{C}_a)$, an abelian variety. In $T^*\mathcal{M}$ the fiber is an open subset of this abelian variety. One checks that the missing part has codimension ≥ 2, so the symplectic form, which is exact, must restrict to 0 on the fibers, completing the proof.

3. Some related systems

Polynomial matrices. One of the earliest appearances of an ACIHS (algebraically completely integrable Hamiltonian system) was in Jacobi's work on the geodesic flow on an ellipsoid (or more generally, on a nonsingular quadric in R^k). Jacobi discovered that this differential equation, taking place on the tangent (= cotangent!) bundle of the ellipsoid, can be integrated explicitly in terms of hyperelliptic theta functions. In our language, the total space of the flow is an ACIHS, fibered by (Zariski open subsets of) hyperelliptic Jacobians. We are essentially in the special case of Proposition 2, where

$$C = P^1, \quad n = 2, \quad K = \mathcal{O}_{P^1}(k).$$

A variant of this system appeared in Mumford's solution [Mu1] of the Schottky problem for hyperelliptic curves.

The extension to all values of n is studied in [B] and, somewhat more analytically, in [AHP] and [AHH]. Beauville considers, for fixed n and k, the space B of polynomials

$$p = y^n + a_1(x)y^{n-1} + \cdots + a_n(x), \quad \deg(a_i) \leq ki \tag{7}$$

in variables x and y. Each p determines an n-sheeted branched cover

$$\widetilde{C}_p \to P^1.$$

The total space is the space of polynomial matrices

$$M := H^0(P^1, \mathrm{End}(\mathcal{O}^{\oplus n}) \otimes \mathcal{O}(d)), \tag{8}$$

the map $h : M \to B$ is the characteristic polynomial, and M_p is the fiber over a given $p \in B$. The result is that, for smooth spectral curves $\widetilde{C}_p$, $\mathrm{PGL}(n)$ acts freely and properly on M_p; the quotient is isomorphic to $J(\widetilde{C}_p) \smallsetminus \Theta$. (In order to obtain the entire $J(\widetilde{C}_p)$, one must allow all pairs (E, ϕ) with E of given degree, say 0. Among those, the ones with $E \approx \mathcal{O}_{P^1}{}^{\oplus n}$ correspond to the open set $J(\widetilde{C}_p) \smallsetminus \Theta$.) This system is an ACIHS, in a slightly weaker sense than before:

instead of a symplectic structure, it has a *Poisson structure*, i.e., a section β of $\wedge^2 T$, such that the **C**-linear sheaf map given by contraction with β

$$\begin{array}{ccc} \mathcal{O} & \to & \mathcal{T} \\ f & \mapsto & df \rfloor \beta \end{array}$$

sends the Poisson bracket of functions to the bracket of vector fields. Any Poisson manifold is naturally foliated, with (locally analytic) symplectic leaves. For a Poisson ACIHS, we want each leaf to inherit a (symplectic) ACIHS, so the symplectic foliation should be pulled back via h from a foliation of the base B.

The result of [BNR] suggests that analogous systems should exist when P^1 is replaced by an arbitrary base curve C. The main point is to construct the Poisson structure. This was achieved by Bottacin [Bn] and Markman [M1]; see Section 6. In the case of the polynomial matrices, though, everything—the commuting vector fields, the Poisson structure, and so on—can be written very explicitly. What makes these explicit results possible is that every vector bundle over P^1 splits. This of course fails in genus > 1, but for elliptic curves the moduli space of vector bundles is still completely understood, so here too the system can be described explicitly, as follows.

For simplicity, consider vector bundles with fixed determinant. When the degree is 0, the moduli space is a projective space P^{n-1} (or, more canonically, the fiber over 0 of the Abel–Jacobi map

$$C^{[n]} \longrightarrow J(C) = C.)$$

The ACIHS that arises is essentially the Treibich–Verdier theory [TV] of elliptic solitons. When, on the other hand, the degree is one (or, more generally, relatively prime to n), the moduli space is a single point; the corresponding system was studied explicitly in [RS].

Reductive groups. In another direction, one can replace the vector bundles by principal G-bundles $\mathcal{G}$ for any reductive group G. Again, there is a moduli space $\mathcal{M}_{G,K}$ parametrizing equivalence classes of semistable K-valued G-Higgs bundles, i.e., pairs $(\mathcal{G}, \phi)$ with $\phi \in K \otimes \mathfrak{ad}(\mathcal{G})$. The Hitchin map goes to

$$B := \bigoplus_i H^0(K^{\otimes d_i}),$$

where the d_i are the degrees of the f_i, a basis for the G-invariant polynomials on the Lie algebra $\mathfrak{g}$:

$$h : (\mathcal{G}, \phi) \longrightarrow (f_i(\phi))_i.$$

When $K = \omega_C$, Hitchin showed [H1] that one still gets a completely integrable system, and that it is algebraically completely integrable for the classical groups $\mathrm{GL}(n), \mathrm{SL}(n), \mathrm{SP}(n), \mathrm{SO}(n)$. The generic fibers are in each case isomorphic (though not quite canonically—one must choose various square roots; see Sections 5.2 and 5.3) to abelian varieties given in terms of the spectral curves $\widetilde{C}$:

$$
(9)\quad \begin{cases}
\mathrm{GL}(n) & \widetilde{C} \text{ has degree } n \text{ over } C, \text{ the AV is } \mathrm{Jac}(\widetilde{C}). \\
\mathrm{SL}(n) & \widetilde{C} \text{ has degree } n \text{ over } C, \text{ the AV is } \mathrm{Prym}(\widetilde{C}/C). \\
\mathrm{SP}(n) & \widetilde{C} \text{ has degree } 2n \text{ over } C \text{ and an involution } x \mapsto -x. \\
 & \text{The map factors through the quotient } \overline{C}. \\
 & \text{The AV is } \mathrm{Prym}(\widetilde{C}/\overline{C}). \\
\mathrm{SO}(n) & \widetilde{C} \text{ has degree } n \text{ and an involution, with:} \\
 & \bullet \text{ a fixed component, when } n \text{ is odd.} \\
 & \bullet \text{ some fixed double points, when } n \text{ is even.} \\
 & \text{One must desingularize } \widetilde{C} \text{ and the quotient } \overline{C}, \text{ and} \\
 & \text{ends up with the Prym of the desingularized double cover.}
\end{cases}
$$

The algebraic complete integrability was verified in [KP1] for the exceptional group G_2. A sketch of the argument for any reductive G is in [BK], and a complete proof was given in [F1]. We will outline a proof in Section 5 below.

Higher dimensions. Finally, a sweeping extension of the notion of Higgs bundle is suggested by the work of Simpson [S]. To him, a Higgs bundle on a projective variety S is a vector bundle (or principal G-bundle ...) E with a symmetric, Ω^1_S-valued endomorphism

$$\phi : E \longrightarrow E \otimes \Omega^1_S.$$

Here *symmetric* means the vanishing of

$$\phi \wedge \phi : E \longrightarrow E \otimes \Omega^2_S,$$

a condition that is obviously vacuous on curves. He constructs a moduli space for such Higgs bundles (satisfying appropriate stability conditions), and establishes diffeomorphisms to corresponding moduli spaces of representations of $\pi_1(S)$ and of connections.

4. Decomposition of spectral Picards

4.1. The question. Let $(\mathcal{G}, \phi)$ be a K-valued principal Higgs bundle on a complex variety S. Each representation

$$\rho : G \longrightarrow \mathrm{Aut}(V)$$

determines an associated K-valued Higgs bundle

$$(\mathcal{V} := \mathcal{G} \times^G V, \quad \rho(\phi)),$$

which in turn determines a spectral cover $\widetilde{S}_V \longrightarrow S$.

The question, raised first in [AvM] when $S = P^1$, is to relate the Picard varieties of the $\widetilde{S}_V$ as V varies, and in particular to find pieces common to all of them. For Adler and van Moerbeke, the motivation was that many evolution differential equations (of Lax type) can be *linearized* on the Jacobians of spectral curves. This means that the 'Liouville tori', which live naturally in the complexified domain of the differential equation (and hence are independent of the representation V) are mapped isogenously to their image in $\mathrm{Pic}(\widetilde{S}_V)$ for each nontrivial V; so one should be able to locate these tori among the pieces that occur in an isogeny decomposition of each of the $\mathrm{Pic}(\widetilde{S}_V)$. There are many specific examples where a pair of abelian varieties constructed from related covers of curves are known to be isomorphic or isogenous, and some of these lead to important identities among theta functions.

EXAMPLE 3. Take $G = \mathrm{SL}(4)$. The standard representation V gives a branched cover $\widetilde{S}_V \longrightarrow S$ of degree 4. On the other hand, the 6-dimensional representation $\wedge^2 V$ (= the standard representation of the isogenous group SO(6)) gives a cover $\overset{\approx}{S} \longrightarrow S$ of degree 6, which factors through an involution

$$\overset{\approx}{S} \longrightarrow \overline{S} \longrightarrow S.$$

One has the isogeny decompositions

$$\mathrm{Pic}\,(\widetilde{S}) \sim \mathrm{Prym}(\widetilde{S}/S) \oplus \mathrm{Pic}\,(S)$$

$$\mathrm{Pic}\,(\overset{\approx}{S}) \sim \mathrm{Prym}(\overset{\approx}{S}/\overline{S}) \oplus \mathrm{Prym}(\overline{S}/S) \oplus \mathrm{Pic}\,(S).$$

It turns out that

$$\mathrm{Prym}(\widetilde{S}/S) \sim \mathrm{Prym}(\overset{\approx}{S}/\overline{S}).$$

For $S = P^1$, this is Recillas' *trigonal construction* [R]. It says that every Jacobian of a trigonal curve is the Prym of a double cover of a tetragonal curve, and vice versa.

EXAMPLE 4. Take $G = \mathrm{SO}(8)$ with its standard 8-dimensional representation V. The spectral cover has degree 8 and factors through an involution, $\overset{\approx}{S} \longrightarrow \overline{S} \longrightarrow S$. The two half-spin representations V_1, V_2 yield similar covers

$$\overset{\approx}{S}_1 \longrightarrow \overline{S}_1 \longrightarrow S, \qquad \overset{\approx}{S}_2 \longrightarrow \overline{S}_2 \longrightarrow S.$$

The *tetragonal construction* [D1] says that the three Pryms of the double covers are isomorphic. (These examples, as well as Pantazis' *bigonal construction* and constructions based on some exceptional groups, are discussed in the context of spectral covers in [K] and [D2].)

It turns out in general that there is indeed a distinguished, Prym-like isogeny component common to all the spectral Picards, on which the solutions to Lax-type differential equations evolve linearly. This was noticed in some cases already in [AvM], and was greatly extended by Kanev's construction of Prym–Tyurin varieties. (He still needs S to be P^1 and the spectral cover to have generic ramification; some of his results apply only to *minuscule representations*.) Various parts of the general story have been worked out recently by a number of authors, based on either of two approaches: one, pursued in [D2, Me, MS], is to decompose everything according to the action of the Weyl group W and to look for common pieces; the other, used in [BK, D3, F1, Sc], relies on the correspondence of spectral data and Higgs bundles. The group-theoretic approach is described in the rest of this section. We take up the second method, known as *abelianization*, in Section 5.

4.2. Decomposition of spectral covers. The decomposition of spectral Picards arises from three sources. First, the spectral cover for a sum of representations is the union of the individual covers $\widetilde{S}_V$. Next, the cover $\widetilde{S}_V$ for an irreducible representation is still the union of subcovers $\widetilde{S}_\lambda$ indexed by weight orbits. And finally, the Picard of $\widetilde{S}_\lambda$ decomposes into Pryms. We start with a few observations about the dependence of the covers themselves on the representation. The decomposition of the Picards is taken up in the next subsection.

Spectral covers. There is an *infinite* collection (of irreducible representations $V := V_\mu$, hence) of spectral covers $\widetilde{S}_V$, which can be parametrized by their highest weights μ in the dominant Weyl chamber $\overline{C}$, or equivalently by the W-orbit of extremal weights, in Λ/W. Here T is a maximal torus in G, $\Lambda := \mathrm{Hom}(T, \mathbf{C}^*)$ is the *weight lattice* (also called *character lattice*) for G, and W is the Weyl group. Each of these $\widetilde{S}_V$ decomposes as the union of its subcovers $\widetilde{S}_\lambda$, parametrizing eigenvalues in a given W-orbit $W\lambda$. (Here λ runs over the weight-orbits in V_μ.)

Parabolic covers. There is a *finite* collection of covers $\widetilde{S}_P$, parametrized by the conjugacy classes in G of parabolic subgroups (or equivalently by arbitrary-dimensional faces F_P of the chamber $\overline{C}$) such that (for general S) each eigenvalue cover $\widetilde{S}_\lambda$ is birational to some parabolic cover $\widetilde{S}_P$, the one whose open face F_P contains λ.

The cameral cover. There is a W-Galois cover $\widetilde{S} \longrightarrow S$ such that each $\widetilde{S}_P$ is isomorphic to $\widetilde{S}/W_P$, where W_P is the Weyl subgroup of W that stabilizes F_P. We call $\widetilde{S}$ the *cameral cover*, since, at least generically, it parametrizes the chambers determined by ϕ (in the duals of the Cartans), or equivalently the Borel subalgebras containing ϕ. This is constructed as follows: There is a morphism $\mathfrak{g} \longrightarrow \mathfrak{t}/W$ sending $g \in \mathfrak{g}$ to the conjugacy class of its semisimple part g_{ss}. (More precisely, this is Spec of the composed ring homomorphism $\mathbf{C}[\mathfrak{t}]^W \overset{\approx}{\leftarrow} \mathbf{C}[\mathfrak{g}]^G \hookrightarrow \mathbf{C}[\mathfrak{g}]$.) Taking fiber product with the quotient map $\mathfrak{t} \longrightarrow \mathfrak{t}/W$, we get the cameral cover $\tilde{\mathfrak{g}}$ of $\mathfrak{g}$. The cameral cover $\widetilde{S} \longrightarrow S$ of a K-valued principal Higgs bundle on S is glued from covers of open subsets in S (on which K and $\mathcal{G}$ are trivialized) that in turn are pullbacks by ϕ of $\tilde{\mathfrak{g}} \longrightarrow \mathfrak{g}$.

4.3. Decomposition of spectral Picards. The decomposition of the Picard varieties of spectral covers can be described as follows:

The cameral Picard. From each isomorphism class of irreducible W-representations, choose an integral representative Λ_i. (This can always be done for Weyl groups.) The group ring $Z[W]$ that acts on $\mathrm{Pic}(\widetilde{S})$ has an isogeny decomposition

$$Z[W] \sim \bigoplus_i \Lambda_i \otimes_Z \Lambda_i^*, \tag{10}$$

which is just the decomposition of the regular representation. There is a corresponding isotypic decomposition

$$\mathrm{Pic}(\widetilde{S}) \sim \bigoplus_i \Lambda_i \otimes_Z \mathrm{Prym}\,\Lambda_i(\widetilde{S}), \tag{11}$$

where

$$\mathrm{Prym}_{\Lambda_i}(\widetilde{S}) := \mathrm{Hom}_W(\Lambda_i, \mathrm{Pic}(\widetilde{S})). \tag{12}$$

Parabolic Picards. There are at least three reasonable ways of obtaining an isogeny decomposition of $\mathrm{Pic}(\widetilde{S}_P)$, for a parabolic subgroup $P \subset G$:

- The 'Hecke' ring Corr_P of correspondences on $\widetilde{S}_P$ over S acts on $\mathrm{Pic}(\widetilde{S}_P)$, so every irreducible integral representation M of Corr_P determines a generalized Prym
$$\mathrm{Hom}_{\mathrm{Corr}_P}(M, \mathrm{Pic}(\widetilde{S}_P)),$$
and we obtain an isotypic decomposition of $\mathrm{Pic}(\widetilde{S}_P)$ as before.
- $\mathrm{Pic}(\widetilde{S}_P)$ maps, with torsion kernel, to $\mathrm{Pic}(\widetilde{S})$, so we obtain a decomposition of the former by intersecting its image with the isotypic components $\Lambda_i \otimes_Z \mathrm{Prym}_{\Lambda_i}(\widetilde{S})$ of the latter.
- Since $\widetilde{S}_P$ is the cover of S *associated* to the W-cover $\widetilde{S}$ via the permutation representation $Z[W_P\backslash W]$ of W, we get an isogeny decomposition of $\mathrm{Pic}(\widetilde{S}_P)$ indexed by the irreducible representations in $Z[W_P\backslash W]$.

It turns out [D2, Section 6] that all three decompositions agree and can be given explicitly as

$$\bigoplus_i M_i \otimes \mathrm{Prym}_{\Lambda_i}(\widetilde{S}) \subset \bigoplus_i \Lambda_i \otimes \mathrm{Prym}_{\Lambda_i}(\widetilde{S}), \qquad M_i := (\Lambda_i)^{W_P}. \tag{13}$$

Spectral Picards. To obtain the decomposition of the Picards of the original covers $\widetilde{S}_V$ or $\widetilde{S}_\lambda$, we need, in addition to the decomposition of $\text{Pic}(\widetilde{S}_P)$, some information on the singularities. These can arise from two separate sources:

Accidental singularities of the $\widetilde{S}_\lambda$. For a sufficiently general Higgs bundle, and for a weight λ in the interior of the face F_P of the Weyl chamber $\overline{C}$, the natural map

$$i_\lambda : \widetilde{S}_P \longrightarrow \widetilde{S}_\lambda$$

is birational. For the *standard* representations of the classical groups of types A_n, B_n or C_n, this *is* an isomorphism. But for general λ it is *not*: In order for i_λ to be an isomorphism, λ must be a multiple of a fundamental weight [D2, Lemma 4.2]. In fact, the list of fundamental weights for which this happens is quite short; for the classical groups we have only: ω_1 for A_n, B_n and C_n, ω_n (the dual representation) for A_n, and ω_2 for B_2. Note that for D_n the list is *empty*. In particular, the covers produced by the standard representation of SO$(2n)$ are singular; this fact, noticed by Hitchin in [H1], explains the need for desingularization in his result (9).

Gluing the $\widetilde{S}_V$. In addition to the singularities of each i_λ, there are the singularities created by the gluing map $\coprod_\lambda \widetilde{S}_\lambda \longrightarrow \widetilde{S}_V$. This makes explicit formulas somewhat simpler in the case, studied by Kanev [K], of *minuscule* representations, i.e., representations whose weights form a single W-orbit. These singularities account, for instance, for the desingularization required in the SO$(2n+1)$ case in (9).

4.4. The distinguished Prym. Combining much of the above, the Adler–van Moerbeke problem of finding a component common to the $\text{Pic}(\widetilde{S}_V)$ for all non-trivial V translates into:

> *Find the irreducible representations* Λ_i *of* W *that occur in* $Z[W_P\backslash W]$ *with positive multiplicity for all proper Weyl subgroups* $W_P\backslash W$.

By Frobenius reciprocity, or (13), this is equivalent to:

> *Find the irreducible representations* Λ_i *of* W *such that for every proper Weyl subgroup* $W_P \subsetneqq W$, *the space of invariants* $M_i := (\Lambda_i)^{W_P}$ *is non-zero.*

One solution is now obvious: the *reflection representation* of W acting on the weight lattice Λ has this property. In fact, Λ^{W_P} in this case is just the face F_P of $\overline{C}$. The corresponding component $\text{Prym}_\Lambda(\widetilde{S})$ is called *the distinguished Prym*. We will see in Section 5 that its points correspond, modulo some corrections, to Higgs bundles.

For the classical groups, this turns out to be the only common component. For G_2 and E_6 it turns out [D2, Section 6] that a second common component exists.

The geometric significance of points in these components is not known. As far as I know, the only component other than the distinguished Prym that has arisen 'in nature' is the one associated to the one-dimensional sign representation of W; see Section 7 and [KP2].

5. Abelianization

5.1. Abstract versus K-valued objects. We want to describe the abelianization procedure in a somewhat abstract setting, as an equivalence between *principal Higgs bundles* and certain *spectral data*. Once we fix a *values* vector bundle K, we obtain an equivalence between *K-valued principal Higgs bundles* and *K-valued spectral data*. Similarly, the choice of a representation V of G will determine an equivalence of *K-valued Higgs bundles* (of a given representation type) with K-valued spectral data.

As our model of a W-cover we take the natural quotient map

$$G/T \longrightarrow G/N$$

and its partial compactification

$$\overline{G/T} \longrightarrow \overline{G/N}. \tag{14}$$

Here $T \subset G$ is a maximal torus, and N is its normalizer in G. The quotient G/N parametrizes maximal tori (= Cartan subalgebras) $\mathfrak{t}$ in $\mathfrak{g}$, while G/T parametrizes pairs $\mathfrak{t} \subset \mathfrak{b}$ with $\mathfrak{b} \subset \mathfrak{g}$ a Borel subalgebra. An element $x \in \mathfrak{g}$ is *regular* if the dimension of its centralizer $\mathfrak{c} \subset \mathfrak{g}$ equals $\dim T$ (= the rank of $\mathfrak{g}$). The partial compactifications $\overline{G/N}$ and $\overline{G/T}$ parametrize regular centralizers $\mathfrak{c}$ and pairs $\mathfrak{c} \subset \mathfrak{b}$, respectively.

In constructing the cameral cover in Section 4.2, we used the W-cover $\mathfrak{t} \longrightarrow \mathfrak{t}/W$ and its pullback cover $\widetilde{\mathfrak{g}} \longrightarrow \mathfrak{g}$. Over the open subset $\mathfrak{g}_{\text{reg}}$ of regular elements, the same cover is obtained by pulling back (14) via the map $\alpha : \mathfrak{g}_{\text{reg}} \longrightarrow \overline{G/N}$ sending an element to its centralizer:

$$\begin{array}{ccccc} \mathfrak{t} & \longleftarrow & \widetilde{\mathfrak{g}}_{\text{reg}} & \longrightarrow & \overline{G/T} \\ \downarrow & & \downarrow & & \downarrow \\ \mathfrak{t}/W & \longleftarrow & \mathfrak{g}_{\text{reg}} & \xrightarrow{\alpha} & \overline{G/N} \end{array} . \tag{15}$$

When working with K-valued objects, it is usually more convenient to work with the left-hand side of (15), i.e., with *eigenvalues*. When working with the abstract objects, this is unavailable, so we are forced to work with the *eigenvectors*, or the right-hand side of (15). Thus:

DEFINITION 5. *An abstract* cameral cover *of S is a finite morphism $\widetilde{S} \longrightarrow S$ with W-action, which locally (étale) in S is a pullback of* (14).

DEFINITION 6. *A K-valued cameral cover (for K a vector bundle on S) consists of a cameral cover $\pi : \widetilde{S} \longrightarrow S$ together with an S-morphism*

$$\widetilde{S} \times \Lambda \longrightarrow \mathbb{K} \tag{16}$$

that is W-invariant (W acts on $\widetilde{S}$ and Λ, hence diagonally on $\widetilde{S} \times \Lambda$) and linear in Λ.

We note that a cameral cover $\widetilde{S}$ determines quotients $\widetilde{S}_P$ for parabolic subgroups $P \subset G$. A K-valued cameral cover determines additionally the $\widetilde{S}_\lambda$ for $\lambda \in \Lambda$, as images in $\mathbb{K}$ of $\widetilde{S} \times \{\lambda\}$. The data of (16) is equivalent to a W-equivariant map $\widetilde{S} \longrightarrow \mathfrak{t} \otimes_{\mathbb{C}} K$.

DEFINITION 7. *A G-principal Higgs bundle on S is a pair $(\mathcal{G}, \mathfrak{c})$ with $\mathcal{G}$ a principal G-bundle and $\mathfrak{c} \subset \mathrm{ad}(\mathcal{G})$ a subbundle of regular centralizers.*

DEFINITION 8. *A K-valued G-principal Higgs bundle consists of $(\mathcal{G}, \mathfrak{c})$ as above, together with a section φ of $\mathfrak{c} \otimes K$.*

A principal Higgs bundle $(\mathcal{G}, \mathfrak{c})$ determines a cameral cover $\widetilde{S} \longrightarrow S$ and a homomorphism $\Lambda \longrightarrow \mathrm{Pic}(\widetilde{S})$. Let F be a parameter space for Higgs bundles with a given $\widetilde{S}$. Each non-zero $\lambda \in \Lambda$ gives a non-trivial map $F \longrightarrow \mathrm{Pic}(\widetilde{S})$. For λ in a face F_P of $\overline{C}$, this factors through $\mathrm{Pic}(\widetilde{S}_P)$. The discussion in Section 4.4 now suggests that F should be given roughly by the distinguished Prym,

$$\mathrm{Hom}_W(\Lambda, \mathrm{Pic}(\widetilde{S})).$$

It turns out that this guess needs two corrections. The first correction involves restricting to a coset of a subgroup; the need for this is visible even in the simplest case where $\widetilde{S}$ is étale over S, so $(\mathcal{G}, \mathfrak{c})$ is everywhere regular and semisimple (i.e., $\mathfrak{c}$ is a bundle of Cartans.) The second correction involves a twist along the ramification of $\widetilde{S}$ over S. We explain these corrections in the next two subsections.

5.2. The regular semisimple case: the shift.

EXAMPLE 9. Fix a smooth projective curve C and a line bundle $K \in \mathrm{Pic}(C)$ such that $K^{\otimes 2} \approx \mathcal{O}_C$. This determines an étale double cover $\pi : \widetilde{C} \longrightarrow C$ with involution i, and homomorphisms

$$\begin{array}{rcccl} \pi^* & : & \mathrm{Pic}(C) & \longrightarrow & \mathrm{Pic}(\widetilde{C}), \\ \mathrm{Nm} & : & \mathrm{Pic}(\widetilde{C}) & \longrightarrow & \mathrm{Pic}(C), \\ i^* & : & \mathrm{Pic}(\widetilde{C}) & \longrightarrow & \mathrm{Pic}(\widetilde{C}), \end{array}$$

satisfying

$$1 + i^* = \pi^* \circ \mathrm{Nm}.$$

For $G = \mathrm{GL}(2)$ we have $\Lambda = Z \oplus Z$, and $W = \mathcal{S}_2$ permutes the summands, so

$$\mathrm{Hom}_W(\Lambda, \mathrm{Pic}(\widetilde{C})) \approx \mathrm{Pic}(\widetilde{C}).$$

And, indeed, the Higgs bundles corresponding to $\widetilde{C}$ are parametrized by $\mathrm{Pic}(\widetilde{C})$: send $L \in \mathrm{Pic}(\widetilde{C})$ to $(\mathcal{G}, \mathrm{c})$, where $\mathcal{G}$ has associated rank-two vector bundle $\mathcal{V} := \pi_* L$, and $\mathrm{c} \subset \mathrm{End}(\mathcal{V})$ is $\pi_* \mathcal{O}_{\widetilde{C}}$.

On the other hand, for $G = \mathrm{SL}(2)$ we have $\Lambda = Z$ and $W = \mathcal{S}_2$ acts by ± 1, so

$$\mathrm{Hom}_W(\Lambda, \mathrm{Pic}(\widetilde{S})) \approx \{L \in \mathrm{Pic}(\widetilde{C}) \mid i^* L \approx L^{-1}\} = \ker(1 + i^*).$$

This group has four connected components. The subgroup $\ker(\mathrm{Nm})$ consists of two of these. The connected component of 0 is the classical Prym variety [Mu2]. Now the Higgs bundles correspond, via the above bijection $L \mapsto \pi_* L$, to

$$\{L \in \mathrm{Pic}(\widetilde{C}) \mid \det(\pi_* L) \approx \mathcal{O}_C\} = \mathrm{Nm}^{-1}(K).$$

Thus they form the *non-zero* coset of the subgroup $\ker(\mathrm{Nm})$. (If we return to a higher-dimensional S, it is possible for K not to be in the image of Nm, so there may be *no* $\mathrm{SL}(2)$-Higgs bundles corresponding to such a cover.)

This example generalizes to all G, as follows. The equivalence classes of extensions

$$1 \longrightarrow T \longrightarrow N' \longrightarrow W \longrightarrow 1$$

(in which the action of W on T is the standard one) are parametrized by the group cohomology $H^2(W, T)$. Here the 0 element corresponds to the semidirect product. The class $[N] \in H^2(W, T)$ of the normalizer N of T in G may be 0, as it is for $G = \mathrm{GL}(n)$, $\mathrm{PGL}(n)$, and $\mathrm{SL}(2n+1)$; or it may not, as for $G = \mathrm{SL}(2n)$.

Assume first, for simplicity, that S and $\widetilde{S}$ are connected and projective. There is then a natural group homomorphism

$$c : \mathrm{Hom}_W(\Lambda, \mathrm{Pic}(\widetilde{S})) \longrightarrow H^2(W, T). \tag{17}$$

Algebraically, this is an edge homomorphism for the Grothendieck spectral sequence of equivariant cohomology, which gives the exact sequence

$$0 \longrightarrow H^1(W, T) \longrightarrow H^1(S, \mathcal{C}) \longrightarrow \mathrm{Hom}_W(\Lambda, \mathrm{Pic}(\widetilde{S})) \xrightarrow{\;c\;} H^2(W, T), \tag{18}$$

where $\mathcal{C} := \widetilde{S} \times_W T$. Geometrically, this expresses a *Mumford group* construction: giving $\mathcal{L} \in \mathrm{Hom}(\Lambda, \mathrm{Pic}(\widetilde{S}))$ is equivalent to giving a principal T-bundle $\mathcal{T}$ over $\widetilde{S}$; for $\mathcal{L} \in \mathrm{Hom}_W(\Lambda, \mathrm{Pic}(\widetilde{S}))$, $c(\mathcal{L})$ is the class in $H^2(W, T)$ of the group N' of automorphisms of $\mathcal{T}$ that commute with the action on $\widetilde{S}$ of some $w \in W$.

To remove the restriction on S and $\widetilde{S}$, we need to replace each occurrence of T in (17, 18) by $\Gamma(\widetilde{S},T)$, the global sections of the trivial bundle on $\widetilde{S}$ with fiber T. The natural map $H^2(W,T) \longrightarrow H^2(W,\Gamma(\widetilde{S},T))$ allows us to think of $[N]$ as an element of $H^2(W,\Gamma(\widetilde{S},T))$.

PROPOSITION 10 ([D3]). *Fix an étale W-cover $\pi : \widetilde{S} \longrightarrow S$. The following data are equivalent:*

(1) *Principal G-Higgs bundles $(\mathcal{G}, \mathfrak{c})$ with cameral cover $\widetilde{S}$.*
(2) *Principal N-bundles $\mathcal{N}$ over S whose quotient by T is $\widetilde{S}$.*
(3) *W-equivariant homomorphisms $\mathcal{L} : \Lambda \longrightarrow \mathrm{Pic}(\widetilde{S})$ with $c(\mathcal{L}) = [N] \in H^2(W,\Gamma(\widetilde{S},T))$.*

We observe that while the shifted objects correspond to Higgs bundles, the unshifted objects

$$\mathcal{L} \in \mathrm{Hom}_W(\Lambda, \mathrm{Pic}(\widetilde{S})), \qquad c(\mathcal{L}) = 0$$

come from the $\mathcal{C}$-torsors in $H^1(S, \mathcal{C})$.

5.3. The regular case: the twist along the ramification.

EXAMPLE 11. Modify Example 9 by letting $K \in \mathrm{Pic}(C)$ be arbitrary, and choose a section b of $K^{\otimes 2}$ that vanishes on a simple divisor $B \subset C$. We get a double cover $\pi : \widetilde{C} \longrightarrow C$ branched along B, ramified along a divisor

$$R \subset \widetilde{C}, \quad \pi(R) = B.$$

Via $L \mapsto \pi_* L$, the Higgs bundles still correspond to

$$\{L \in \mathrm{Pic}(\widetilde{C}) \mid \det(\pi_* L) \approx \mathcal{O}_C\} = \mathrm{Nm}^{-1}(K).$$

But this is no longer in $\mathrm{Hom}_W(\Lambda, \mathrm{Pic}(\widetilde{S}))$; rather, the line bundles in question satisfy

$$i^* L \approx L^{-1}(R). \tag{19}$$

For arbitrary G, let Φ denote the root system and Φ^+ the set of positive roots. There is a decomposition

$$\overline{G/T} \setminus G/T = \bigcup_{\alpha \in \Phi^+} R_\alpha$$

of the boundary into components, with R_α the fixed locus of the reflection σ_α in α. (Via (15), these correspond to the complexified walls in $\mathfrak{t}$.) Thus each cameral cover $\widetilde{S} \longrightarrow S$ comes with a natural set of (Cartier) *ramification divisors*, which we still denote R_α, for $\alpha \in \Phi^+$.

For $w \in W$, set

$$F_w := \{\alpha \in \Phi^+ \mid w^{-1}\alpha \in \Phi^-\} = \Phi^+ \cap w\Phi^-,$$

and choose a W-invariant form $\langle , \rangle$ on Λ. We consider the variety

$$\mathrm{Hom}_{W,R}(\Lambda, \mathrm{Pic}(\widetilde{S}))$$

of R-twisted W-equivariant homomorphisms, i.e., homomorphisms $\mathcal{L}$ satisfying

$$w^*\mathcal{L}(\lambda) \approx \mathcal{L}(w\lambda)\left(\sum_{\alpha \in F_w} \frac{\langle -2\alpha, w\lambda\rangle}{\langle \alpha, \alpha\rangle} R_\alpha\right), \qquad \lambda \in \Lambda, \quad w \in W. \tag{20}$$

This turns out to be the correct analogue of (19). (For example if $w = \sigma_\alpha$, is a reflection, F_w is $\{\alpha\}$, so this gives

$$w^*\mathcal{L}(\lambda) \approx \mathcal{L}(w\lambda)\left(\frac{\langle \alpha, 2\lambda\rangle}{\langle \alpha, \alpha\rangle} R_\alpha\right),$$

which specializes to (19).) As before, there is a class map

$$c : \mathrm{Hom}_{W,R}(\Lambda, \mathrm{Pic}(\widetilde{S})) \longrightarrow H^2(W,\ \Gamma(\widetilde{S}, T)) \tag{21}$$

that can be described via a Mumford-group construction.

To understand this twist, consider the formal object

$$\begin{aligned} \tfrac{1}{2}\,\mathrm{Ram} :\ \Lambda &\longrightarrow Q \otimes \mathrm{Pic}\,\widetilde{S}, \\ \lambda &\longmapsto \sum_{(\alpha \in \Phi^+)} \frac{\langle \alpha, \lambda\rangle}{\langle \alpha, \alpha\rangle} R_\alpha. \end{aligned}$$

In an obvious sense, a principal T-bundle $\mathcal{T}$ on $\widetilde{S}$ (or a homomorphism $\mathcal{L} : \Lambda \longrightarrow \mathrm{Pic}(\widetilde{S})$) is R-twisted W-equivariant if and only if $\mathcal{T}(-\frac{1}{2}\,\mathrm{Ram})$ is W-equivariant, i.e., if $\mathcal{T}$ and $\frac{1}{2}\,\mathrm{Ram}$ transform the same way under W. The problem with this is that $\frac{1}{2}\,\mathrm{Ram}$ itself does not make sense as a T-bundle, because the coefficients $\langle \alpha, \lambda\rangle / \langle \alpha, \alpha\rangle$ are not integers. (This argument shows that if $\mathrm{Hom}_{W,R}(\Lambda, \mathrm{Pic}(\widetilde{S}))$ is non-empty, it is a torsor over the untwisted $\mathrm{Hom}_W(\Lambda, \mathrm{Pic}(\widetilde{S}))$.)

THEOREM 12 ([D3]). *For a cameral cover $\widetilde{S} \longrightarrow S$, the following data are equivalent:*

(1) *G-principal Higgs bundles with cameral cover $\widetilde{S}$.*
(2) *R-twisted W-equivariant homomorphisms $\mathcal{L} \in c^{-1}([N])$.*

The theorem has an essentially local nature, as there is no requirement that S be, say, projective. We also do not need the condition of generic behavior near the ramification, which appears in [F1, Me, Sc]. Thus we may consider an extreme case where $\widetilde{S}$ is 'everywhere ramified':

EXAMPLE 13. In Example 11, take the section $b = 0$. The resulting cover $\widetilde{C}$ is a 'ribbon', or length-2 non-reduced structure on C: it is the length-2 neighborhood of C in $\mathbb{K}$. The SL(2)-Higgs bundles $(\mathcal{G}, \mathrm{c})$ for this $\widetilde{C}$ have an everywhere nilpotent c, so the vector bundle $\mathcal{V} := \mathcal{G} \times^{\mathrm{SL}(2)} V \approx \pi_* L$ (where V is the standard two-dimensional representation) fits in an exact sequence

$$0 \longrightarrow \mathcal{S} \longrightarrow \mathcal{V} \longrightarrow \mathcal{Q} \longrightarrow 0$$

with $\mathcal{S} \otimes K \approx \mathcal{Q}$. Such data are specified by the line bundle $\mathcal{Q}$, satisfying $\mathcal{Q}^{\otimes 2} \approx K$, and an extension class in $\mathrm{Ext}^1(\mathcal{Q}, \mathcal{S}) \approx H^1(K^{-1})$. The kernel of the restriction map $\mathrm{Pic}(\widetilde{C}) \longrightarrow \mathrm{Pic}(C)$ is also given by $H^1(K^{-1})$ (use the exact sequence $0 \longrightarrow K^{-1} \longrightarrow \pi_* \mathcal{O}^{\times}_{\widetilde{C}} \longrightarrow \mathcal{O}^{\times}_{C} \longrightarrow 0$), and the R-twist produces the required square roots of K. (For more details on the nilpotent locus, see [L] and [DEL].)

5.4. Adding values and representations. Fix a vector bundle K, and consider the moduli space $\mathcal{M}_{S,G,K}$ of K-valued G-principal Higgs bundles on S. (It can be constructed as in Simpson's [S], even though the objects we need to parametrize are slightly different from his. In this subsection we outline a direct construction.) It comes with a Hitchin map

$$h : \mathcal{M}_{S,G,K} \longrightarrow B_K \tag{22}$$

where $B := B_K$ parametrizes all possible Hitchin data. Theorem 12 gives a precise description of the fibers of this map, independent of the values bundle K. This leaves us with the relatively minor task of describing, for each K, the corresponding base, that is, the closed subvariety B_s of B parametrizing *split* Hitchin data, or K-valued cameral covers. The point is that Higgs bundles satisfy a symmetry condition, which in Simpson's setup is

$$\varphi \wedge \varphi = 0,$$

and is built into our Definition 7 through the assumption that c is regular, hence abelian. Since commuting operators have common eigenvectors, this gives a splitness condition on the Hitchin data, which we describe below. (When K is a line bundle, the condition is vacuous, $B_s = B$.) The upshot is:

LEMMA 14. *The following data are equivalent:*
(a) *A K-valued cameral cover of S.*
(b) *A split, graded homomorphism $R^{\bullet} \longrightarrow Sym^{\bullet}K$.*
(c) *A split Hitchin datum $b \in B_s$.*

Here $R^{\bullet}$ is the graded ring of W-invariant polynomials on $\mathfrak{t}$:

(23) $$R^{\cdot} := (\mathrm{Sym}^{\cdot}\mathfrak{t}^*)^W \approx \mathbf{C}[\sigma_1, \ldots, \sigma_l], \qquad \deg(\sigma_i) = d_i$$

where $l := \mathrm{Rank}(\mathfrak{g})$ and the σ_i form a basis for the W-invariant polynomials. The Hitchin base is the vector space

$$B := B_K := \oplus_{i=1}^{l} H^0(S, Sym^{d_i} K) \approx \mathrm{Hom}(R^{\cdot}, \mathrm{Sym}^{\cdot} K).$$

For each $\lambda \in \Lambda$ (or $\lambda \in \mathfrak{t}^*$, for that matter), the expression

(24) $$q_\lambda(x,t) := \prod_{w \in W} (x - w\lambda(t)), \qquad t \in \mathfrak{t},$$

in an indeterminate x is W-invariant (as a function of t), so it defines an element $q_\lambda(x) \in R^{\cdot}[x]$. A Hitchin datum $b \in B \approx \mathrm{Hom}(R^{\cdot}, \mathrm{Sym}^{\cdot} K)$ sends this to

$$q_{\lambda,b}(x) \in \mathrm{Sym}^{\cdot}(K)[x].$$

We say that b is *split* if, at each point of S and for each λ, the polynomial $q_{\lambda,b}(x)$ factors completely, into terms linear in x.

We note that, for λ in the interior of C (the positive Weyl chamber), $q_{\lambda,b}$ gives the equation in $\mathbb{K}$ of the spectral cover $\widetilde{S}_\lambda$ of Section (4.2): $q_{\lambda,b}$ gives a morphism $\mathbb{K} \longrightarrow \mathrm{Sym}^N \mathbb{K}$, where $N := \#W$, and $\widetilde{S}_\lambda$ is the inverse image of the zero-section. (When λ is in a face F_P of $\overline{C}$, we define analogous polynomials $q_\lambda^P(x,t)$ and $q_{\lambda,b}^P(x)$ by taking the product in (24) to be over $w \in W_P \backslash W$. These give the reduced equations in this case, and q_λ is an appropriate power.)

Over B_s there is a universal K-valued cameral cover

$$\widetilde{S} \longrightarrow B_s$$

with ramification divisor $R \subset \widetilde{S}$. From the relative Picard, $\mathrm{Pic}(\widetilde{S}/B_s)$, we concoct the relative N-shifted, R-twisted Prym

$$\mathrm{Prym}_{\Lambda,R}(\widetilde{S}/B_s).$$

By Theorem 12, this can then be considered as a parameter space $\mathcal{M}_{S,G,K}$ for all K-valued G-principal Higgs bundles on S. (Recall that our objects are assumed to be everywhere *regular*!) It comes with a 'Hitchin map', namely the projection to B_s, and the fibers corresponding to smooth projective $\widetilde{S}$ are abelian varieties. When S is a smooth, projective curve, we recover this way the algebraic complete integrability of Hitchin's system and its generalizations.

6. Symplectic and Poisson structures

The total space of Hitchin's original system is a cotangent bundle, hence has a natural symplectic structure. For the polynomial matrix systems of [B] and [AHH] there is a natural Poisson structure, which one writes down explicitly.

In [Bn] and [M1], this result is extended to the systems $\mathcal{M}_{C,K}$ of K-valued GL(n) Higgs bundles on C, when $K \approx \omega_C(D)$ for an effective divisor D on C. There is a general-nonsense pairing on the cotangent spaces, so the point is to check that this pairing is 'closed', that is, satisfies the identity required for a Poisson structure. Bottacin does this by an explicit computation along the lines of [B]. Markman's idea is to consider the moduli space $\mathcal{M}_D$ of stable vector bundles on C with level-D structure. He realizes an open subset $\mathcal{M}^0_{C,K}$ of $\mathcal{M}_{C,K}$, parametrizing Higgs bundles whose covers are nice, as a quotient (by an action of the level group) of $T^*\mathcal{M}_D$, so the natural symplectic form on $T^*\mathcal{M}_D$ descends to a Poisson structure on $\mathcal{M}^0_{C,K}$. This is identified with the general-nonsense form (wherever both exist), proving its closedness.

In [Muk], Mukai constructs a symplectic structure on the moduli space of simple sheaves on a K3 surface S. Given a curve $C \subset S$, one can consider the moduli of sheaves having the numerical invariants of a line bundle on a curve in the linear system $|nC|$ on S. This has a support map to the projective space $|nC|$, which turns it into an ACIHS. This system specializes, by a 'degeneration to the normal cone' argument [DEL] to Hitchin's, allowing translation of various results about Hitchin's system (such as Laumon's description of the nilpotent cone [L]) to Mukai's.

In higher dimensions, the moduli space $\mathcal{M}$ of Ω^1-valued Higgs bundles carries a natural symplectic structure [S]. (Corlette points out in [C] that certain components of an open subet in $\mathcal{M}$ can be described as cotangent bundles.) It is not clear at the moment exactly when one should expect to have an ACIHS, with symplectic, Poisson or quasi-symplectic structure, on the moduli spaces of K-valued Higgs bundles for higher-dimensional S, arbitrary G, and arbitrary vector bundle K. A beautiful new idea [M2] is that Mukai's results extend to the moduli of those sheaves on a (symplectic, Poisson or quasi-symplectic) variety X whose support in X is *Lagrangian*. Again, there is a general-nonsense pairing. At points where the support is non-singular projective, this can be identified with another, more geometric pairing, constructed using the *cubic condition* of [DM1], which is known to satisfy the closedness requirement. This approach is quite powerful, as it includes many non-linear examples such as Mukai's, in addition to the line-bundle valued spectral systems of [Bn, M1] and also Simpson's Ω^1-valued GL(n)-Higgs bundles: just take $X := T^*S \xrightarrow{\pi} S$, with its natural symplectic form, and the support in X to be proper over S of degree n; such sheaves correspond to Higgs bundles by π_*.

The structure group $\mathrm{GL}(n)$ can of course be replaced by an arbitrary reductive group G. Using Theorem 12, this yields (in the analogous cases) a Poisson structure on the Higgs moduli space $\mathcal{M}_{S,G,K}$ described at the end of the previous section. The fibers of the generalized Hitchin map are Lagrangian with respect to this structure. Along the lines of our general approach, the necessary modifications are clear: π_* is replaced by the equivalence of Theorem 12. One thus considers only Lagrangian supports that retain a W-action, and only *equivariant* sheaves on them (with the numerical invariants of a line bundle). These two restrictions are symplectically dual, so the moduli space of Lagrangian sheaves with these invariance properties is a symplectic (respectively, Poisson) subspace of the total moduli space, and the fibers of the Hitchin map are Lagrangian as expected.

A more detailed review of the ACIHS aspects of Higgs bundles will appear in [DM2].

7. Some applications and problems

Some applications. In [H1], Hitchin used his integrable system to compute several cohomology groups of the moduli space $\mathcal{SM}$ (of rank-two, fixed odd determinant vector bundles on a curve C) with coefficients in symmetric powers of its tangent sheaf $\mathcal{T}$. The point is that the symmetric algebra $\mathrm{Sym}^{\bullet}\mathcal{T}$ is the direct image of $\mathcal{O}_{T^*\mathcal{SM}}$, and sections of the latter all pull back via the Hitchin map h from functions on the base B, since the fibers of h are open subsets in abelian varieties, and the missing locus has codimension ≥ 2. Hitchin's system is used in [BNR] to compute a couple of 'Verlinde numbers' for $\mathrm{GL}(n)$, namely the dimensions

$$h^0(\mathcal{M}, \Theta) = 1, \qquad h^0(\mathcal{SM}, \Theta) = n^g.$$

These results are now subsumed in the general Verlinde formulas; see [F2], [BL], and other references therein.

A pretty application of spectral covers was obtained by Katzarkov and Pantev [KP2]. Let S be a smooth, projective, complex variety, and $\rho : \pi_1(S) \longrightarrow G$ a Zariski dense representation into a simple G (over C). Assume that the Ω^1-valued Higgs bundle $(\mathcal{V}, \phi)$ associated to ρ by Simpson is (regular and) generically semisimple, so the cameral cover is reduced. Among other things, they show that ρ factors through a representation of an orbicurve if and only if the non-standard component $\mathrm{Prym}_\epsilon(\widetilde{S})$ is non-zero, where ϵ is the one-dimensional sign representation of W. (In a sense, this is the opposite of $\mathrm{Prym}_\Lambda(\widetilde{S})$: while $\mathrm{Prym}_\Lambda(\widetilde{S})$ is common to $\mathrm{Pic}(\widetilde{S}_P)$ for all proper Weyl subgroups, $\mathrm{Prym}_\epsilon(\widetilde{S})$ occurs in none except for the full cameral Picard.)

Another application is in [KoP]: the moduli spaces of $\mathrm{SL}(n)$- or $\mathrm{GL}(n)$-stable bundles on a curve have certain obvious automorphisms, coming from tensoring with line bundles on the curve, from inversion, or from automorphisms of the curve. Kouvidakis and Pantev use the dominant direct-image maps from spectral Picards and Pryms to the moduli spaces to show that there are no further, unexpected automorphisms. This then leads to a 'non-abelian Torelli theorem', stating that a curve is determined by the isomorphism class of the moduli space of bundles on it.

Compatibility? Hitchin's construction [H2] of the projectively flat connection on the vector bundle of non-abelian theta functions over the moduli space of curves does not really use much about spectral covers. Nor do other constructions of Faltings [F1] and Witten et al. [APW]. Hitchin's work suggests that the 'right' approach should be based on comparison of the non-abelian connection near a curve C with the abelian connection for standard theta functions on spectral covers $\widetilde{C}$ of C. One conjecture concerning the possible relationship between these connections appears in [A], and some related versions have been attempted by several people, so far in vain. What's missing is a compatibility statement between the actions of the two connections on pulled-back sections. If the expected compatibility turns out to hold, it would give another proof of the projective flatness. It should also imply projective finiteness and projective unitarity of monodromy for the non-abelian thetas, and may or may not bring us closer to a 'finite-dimensional' proof of Faltings' theorem (= the former Verlinde conjecture).

Irregulars? The Higgs bundles we consider in this survey are assumed to be everywhere regular. This is a reasonable assumption for line-bundle valued Higgs bundles on a curve or surface, but *not* in dim ≥ 3. This is because the complement of $\mathfrak{g}_{\mathrm{reg}}$ has codimension 3 in $\mathfrak{g}$. The source of the difficulty is that the analogue of (15) fails over $\mathfrak{g}$. There are two candidates for the universal cameral cover: $\widetilde{\mathfrak{g}}$, defined by the left-hand side of (15), is finite over $\mathfrak{g}$ with W action, but does not have a family of line bundles parametrized by Λ. These live on $\widetilde{\widetilde{\mathfrak{g}}}$, the object defined by the right-hand side, which parametrizes pairs

$$(x, \mathfrak{b}), \qquad x \in \mathfrak{b} \subset \mathfrak{g}.$$

This suggests that the right way to analyze irregular Higgs bundles may involve spectral data consisting of a tower

$$\widetilde{\widetilde{S}} \xrightarrow{\sigma} \widetilde{S} \longrightarrow S$$

together with a homomorphism $\mathcal{L} : \Lambda \longrightarrow \mathrm{Pic}(\widetilde{\widetilde{S}})$ such that the collection of sheaves

$$\sigma_*(\mathcal{L}(\lambda)), \qquad \lambda \in \Lambda,$$

on $\widetilde{S}$ is R-twisted W-equivariant in an appropriate sense. As a first step, one may wish to understand the direct images $R^i\sigma_*(\mathcal{L}(\lambda))$ and in particular the cohomologies $H^i(F, \mathcal{L}(\lambda))$, where F, usually called a *Springer fiber*, is a fiber of σ. For regular x, this fiber is a single point. For $x = 0$, the fiber is all of G/B, so the fiber cohomology is given by the Borel–Weil–Bott theorem. The question may thus be considered as a desired extension of BWB to general Springer fibers.

References

[AHH] M.R. Adams, J. Harnad and J. Hurtubise, *Isospectral Hamiltonian flows in finite and infinite dimensions II: Integration of flows*, Comm. Math. Phys. **134** (1990), 555–585.

[A] M. Atiyah, *The geometry and physics of knots*, Cambridge University Press, 1990.

[AHP] M. R. Adams, J. Harnad and E. Previato, *Isospectral Hamiltonian flows in finite and infinite dimensions* I: *Generalized Moser systems and moment maps into loop algebras*, Comm. Math. Phys. **117** (1988), 451–500.

[AvM] M. Adler and P. van Moerbeke, *Completely integrable systems, Euclidean Lie algebras, and curves*, Adv. Math. **38** (1980), 267–379.

[APW] S. Axelrod, S. della Pietra and E. Witten: *Geometric quantization of Chern–Simons gauge theory*, J. Diff. Geom. **33** (1991), 787–902.

[B] A. Beauville, *Jacobiennes des courbes spectrales et systèmes hamiltoniens complètement intégrables*, Acta Math. **164** (1990), 211–235.

[BK] A. Beilinson and D. Kazhdan, *Flat projective connections*, unpublished (1990).

[BL] A. Beauville and Y. Laszlo, *Conformal blocks and generalized theta functions*, Comm. Math. Phys. (to appear).

[Bn] F. Bottacin, *Symplectic geometry on moduli spaces of stable pairs*, Thesis, Orsay, 1992.

[BNR] A. Beauville, M. S. Narasimhan and S. Ramanan, *Spectral curves and the generalized theta divisor*, J. Reine Angew. Math. **398** (1989), 169–179.

[C] K. Corlette, *Non abelian Hodge theory*, Proc. Sympos. Pure Math. **54.2** (1993), 125–144.

[D1] R. Donagi, *The tetragonal construction*, Bull. Amer. Math. Soc. (N.S.) **4** (1981), 181–185.

[D2] ———, *Decomposition of spectral covers*, Journées de Geometrie Algebrique D'Orsay, Astérisque **218** (1993), 145–175.

[D3] ———, *Abelianization of Higgs bundles*, preprint.

[DEL] R. Donagi, L. Ein and R. Lazarsfeld, *A non-linear deformation of the Hitchin dynamical system*, preprint.

[DM1] R. Donagi and E. Markman, *Cubics, integrable systems, and Calabi–Yau threefolds*, Proc. 1993 Hirzebruch Conf. (to appear); available electronically as Alg. Geom. eprint no. 9408004.

[DM2] ———, *Spectral curves, algebraically completely integrable Hamiltonian systems, and moduli of bundles*, 1993 CIME lecture notes, Lecture Notes in Math. (to appear).

[F1] G. Faltings, *Stable G-bundles and Projective Connections*, Jour. Alg. Geo **2** (1993), 507–568.

[F2] ———, *A proof of the Verlinde formula*, J. Algebraic Geom. (to appear).

[H1] N.J. Hitchin, *Stable bundles and integrable systems*, Duke Math. J. **54** (1987), 91–114.

[H2] ———, *Flat connections and geometric quantization*, Comm. Math. Phys.**131** (1990), 347–380.

[K] V. Kanev, *Spectral curves, simple Lie algebras and Prym–Tjurin varieties*, Proc. Sympos. Pure Math. **49** (1989), Part I, 627–645.

[KoP] A. Kouvidakis and T. Pantev, *Automorphisms of the moduli spaces of stable bundles on a curve*, Math. Ann. (to appear).

[KP1] L. Katzarkov and T. Pantev, *Stable G_2 bundles and algebrically completely integrable systems*, Comp. Math.**92** (1994), 43–60.

[KP2] ———, *Representations of fundamental groups whose Higgs bundles are pullbacks*, J. Differential Geom. **39** (1994), 103–121.

[L] G. Laumon, *Un analogue global du cône nilpotent*, Duke Math. J. **57** (1988), 647–671.

[M1] E. Markman, *Spectral curves and integrable systems*, Comp. Math. **93** (1994), 225–290.

[M2] ———, *Symplectic structure on the moduli space of Lagrangian sheaves*, In preparation.

[Me] J. Y. Merindol, *Varietes de Prym d'un revêtement galoisien*, preprint (1993).

[MS] A. McDaniel and L. Smolinsky, *A Lie theoretic Galois Theory for the spectral curves of an integrable system* II, preprint (1994).

[Mu1] D. Mumford, *Tata Lectures on Theta* II, PM **43** (1984), Birkhäuser.

[Mu2] ———, *Prym varieties* I, Contribution to analysis, Acad. Press (1974), 325–350.

[Muk] S. Mukai, *Symplectic structure of the moduli space of simple sheaves on an abelian or K3 surface*, Invent. Math. **77** (1984), 101–116.

[R] S. Recillas, *Jacobians of curves with g^1_4's are Pryms of trigonal curves*, Bol. Soc. Mat. Mexicana **19** (1974) no.1.

[RS] A. G. Reyman and M. A. Semenov-Tyan-Shansky, *Group theoretical methods in the theory of finite dimensional integrable systems*, Dynamical systems 7, V. I. and Arnol'd, S. P. Novikov (eds.) (EMS, vol. **16**, pp. 119–193) 1987 (Russian).

[S] C. Simpson, *Moduli of representations of the fundamental group of a smooth projective variety*, preprint, Princeton University (1989).

[Sc] R. Scognamillo, *Prym–Tjurin varieties and the Hitchin map*, preprint (1993).

[TV] A. Treibich and J. L. Verdier, *Solitons Elliptiques*, The Grothendieck Festschrift III, Birkhäuser (1990), 437–480.

RON DONAGI
DEPARTMENT OF MATHEMATICS
UNIVERSITY OF PENNSYLVNIA
PHILADELPHIA, PA 19104-6395

E-mail address: donagi@math.upenn.edu

Complex Algebraic Geometry
MSRI Publications
Volume **28**, 1995

Adjoint Linear Systems

LAWRENCE EIN

ABSTRACT. This note describes some of the recent methods and effective results in the study of pluricanonical and adjoint linear systems on higher-dimensional varieties. We describe an algebraic construction for the multiplier ideals and we use it to give a simple proof for the Reider theorem.

§1. The purpose of this note is to survey some of the recent results on pluricanonical and adjoint linear systems on algebraic varieties. Let A be a nef and big divisor on a smooth projective variety X. We would like to study the linear system $|K_X + A|$. For X a curve, it is well known that if $\deg A \geq 2$, the linear system $|K_X + A|$ is free, and if $\deg A \geq 3$, then $K_X + A$ is very ample. The canonical linear system $|K_X|$ is very ample if and only if X is not a hyperelliptic curve. From the theory of curves, we know that the properties of these linear systems are closely related to the geometry of the curve. It is natural that one would like to obtain similar numerical criteria for freeness and very ampleness for adjoint linear systems on a higher-dimensional variety, and study their geometric properties.

Many results and ideas in this note are based on my joint work with R. Lazarsfeld. I would like to thank him for sharing with me many of his ideas.

For surfaces of general type, Kodaira [Kod], Bombieri [Bmb] and many others have studied the behavior of the pluricanonical maps. A few years ago, Reider obtained the following elegant unifying result for adjoint linear systems on surfaces, by using the famous Bogomolov's instability criterion for rank-two bundles [Rdr].

Partially supported by NSF Grant 93-02512

THEOREM 1 (Reider). *Let A be a nef and big divisor on a smooth projective surface X, and let p be a given point in X.*

(a) *Assume that $A^2 \geq 5$ and $A \cdot C \geq 2$ for all curves C in X through p. Then the linear system $|K_X + A|$ is free at p.*
(b) *Assume that $A^2 \geq 10$ and $A \cdot C \geq 3$ for all curves C in X. Then $|K_X + A|$ is very ample.*

Naturally one knows a lot less about general adjoint linear systems on higher-dimensional varieties. Using the Kawamata–Viehweg vanishing theorem and an ingenious use of singular hypersurfaces, Kawamata found the following basic base-point free result that is one of the key steps in the higher-dimensional minimal model program.

THEOREM 2 (Kawamata [KMM]). *Let A be a nef and big divisor on a projective variety with only canonical singularities. Assume that $K_X + A$ is nef and Cartier. Then $|m(K_X + A)|$ is free for all $m \gg 0$.*

REMARK. Kawamata has first proved this in the case that X is a threefold. He also observed that the proof will work in higher dimensions, if one can show certain linear systems are non-empty. Then Shokurov [S] proved the necessary non-vanishing theorem.

Let A be an ample line bundle on a smooth projective n-fold. Mori proved that $K_X + mA$ is nef if $m \geq n + 1$. As for the behavior of the adjoint linear systems, Fujita made the following famous conjectures.

FUJITA'S CONJECTURES [Fuj 1].

(a) *If $m \geq n + 1$, the linear system $|K_X + mA|$ is free.*
(b) *If $m \geq n + 2$, the linear system $|K_X + mA|$ is very ample.*
(c) *If K_X is nef and big and $m \geq n + 2$, then $|mK_X|$ is free.*

REMARKS. (a) By considering the case that X is the projective space and A is the tautological line bundle on the projective space, one sees that the conjectures (a) and (b) are optimal.
(b) For surfaces, one can show that Fujita's conjectures follow from Rieder's theorem easily.

In a real breakthrough, Demailly [Dem] obtained the following effective results in higher dimensions by using powerful analytic tools such as Calabi–Yau's theorem and Nadel's vanishing theorem.

THEOREM 3 (Demailly). *Let B be a nef and big divisor on a smooth projective n-fold X and let p be a point in X. Let T_X denote the tangent sheaf of X. Assume that there is a given nonnegative number m such that*

$$T_X \otimes \mathcal{O}_X(mB)$$

is nef. Then there are n positive constants, $c_1, c_2, \ldots, c_n$, expressible explicitly as functions of n and m alone, with the following property. Suppose that the divisor B satisfies the following numerical conditions. For $k = 1, 2, \ldots, n$,

$$B^k \cdot Y > c_k \quad \text{for all } k\text{-dimensional subvarieties } Y \quad \text{such that} \quad p \in Y.$$

Then the linear system $|K_X + B|$ is free at p.

REMARK. The explicit constants in the above theorem are quite large and are unlikely to be optimal. Demailly has also found results for separating points and separating higher-order jets. Using an iteration argument, he obtains the following explicit result.

COROLLARY 4. *Let A be an ample divisor on a smooth projective n-fold X. Then $2K_X + 12n^n A$ is very ample.*

Together with Lazarsfeld, we begin our study of these questions by observing that one can use the cohomological techniques developed by Kawamata, Reid, Shokurov and others to give a fairly simple proof for Reider's theorem. Then Kollár showed that these techniques give an algebraic proof for the following effective results [Kol1].

THEOREM 5 (Kollár). (a) *Let A be a nef and big divisor on a smooth projective variety X. Assume that $K_X + A$ is nef. Then the linear system*

$$|m(K_X + A)|$$

is free for $m > 2\,(n+2)!\,(n+1)$.

(b) *$h^0(K_X + mA) \neq 0$ if $m > \binom{n+1}{2}$.*

Although these effective results are very interesting, the constants involved are very large. It is natural to ask whether one can find results similar to Reider's theorem for higher-dimensional varieties. For threefolds, Lazarsfeld and I were able to find the following results [EL1].

THEOREM 6 (Ein and Lazarsfeld). *Let X be a smooth projective threefold.*

(a) *Let A be an ample divisor on X. Then $|K_X + 4A|$ is free.*
(b) *If X is a smooth minimal threefold of general type, $|7K_X|$ is free.*
(c) *Let B be a nef and big divisor on X and p be a given point on X. Assume that*

$$B^3 \geq 92,$$
$$B^2 \cdot S \geq 7 \quad \text{for all surfaces } S \text{ in } X \text{ such that } p \in S,$$
$$B \cdot C \geq 3 \quad \text{for all curves in } X \text{ such that } p \in C.$$

Then $|K_X + B|$ is free at p.

REMARK. Theorem 4(a) is the special case of Fujita's conjecture for threefolds. Also, Benveniste had already shown earlier [Ben] that $|mK_X|$ is free when K_X is nef and big and $m \geq 34$.

§2. In this section, we will try to demonstrate the cohomological techniques used in the above theorem by giving a proof for the Reider theorem. We begin by considering the following general set-up.

Let X be a smooth projective n-fold and let p be a point in X. Let D be an effective $\mathbf{Q}$-divisor on X that contains p. We will assume that

$$\mathrm{Mult}_p(D) = q \geq n.$$

DEFINITION. We say that p is an almost isolated singularity of D, if

$$\mathrm{Mult}_x(D) < 1$$

for all points x in a punctured neighborhood of p.

The following result relates singular hypersurfaces to the freeness of the adjoint linear systems. We have learned the result from a lecture of Siu. The algebraic proof given here is based on the construction and properties of multiplier ideals given by Esnault and Viehweg in their lectures on vanishing theorems [EV, §7].

LEMMA 7. *Let X be a smooth projective variety and p be a point in X. Suppose that there is an effective $\mathbf{Q}$-divisor D such that* $\mathrm{Mult}_p(D) \geq n$ *and p is an almost isolated singular point of D. Let B be a divisor such that $B - D$ is nef and big. Then the linear system $|K_X + B|$ is free at p.*

SKETCH OF THE PROOF. Let $f : Y \longrightarrow X$ be an embedded resolution for the divisor D. We consider the $\mathbf{Q}$-divisor

$$f^*D = [f^*D] + \Delta.$$

We also write

$$K_{Y/X} - [f^*D] = P - N,$$

where P and N are effective divisors with no common components. Observe that all the components of P are f-exceptional. It follows that

$$f_*(\mathcal{O}_Y(P - N)) = f_*(\mathcal{O}_Y(-N)) = J_Z,$$

where J_Z is the ideal sheaf of a closed subscheme Z. This is the multiplier ideal associated to the $\mathbf{Q}$-divisor D as defined in [EV, §7.4]. By our assumption, p is an almost isolated singular point of D. It follows from Proposition 7.7 in [EV] that in a neighborhood of p the ideal J_Z is a nontrivial ideal of a zero-dimensional closed subscheme supported at p. Now

$$f^*(K_X + B) + P - N \equiv K_Y + f^*(B - D) + \Delta.$$

By the Kawamata–Viehweg vanishing theorem and the Leray spectral sequence, we conclude that

(a) $R^i f_*(\mathcal{O}_Y(P-N)) = 0 \quad$ for $i > 0$;
(b) $f_*\mathcal{O}_Y(P-N) = J_z$;
(c) $H^i(\mathcal{O}_X(K_X+B) \otimes J_z) = 0 \quad$ for $i > 0$.

Therefore the restriction map

$$H^0(\mathcal{O}_X(K_X+B)) \longrightarrow H^0(\mathcal{O}_X(K_X+B) \mid Z)$$

is surjective. Since Z is zero-dimensional at p, this implies that the linear system $|K_X+B|$ is free at p.

Now we'll apply Lemma 7 to give a proof for Reider's theorem (Theorem 1).

SKETCH OF THE PROOF OF THEOREM 1. By the Riemann–Roch theorem, there is a divisor $D \in |kB|$, for $k \gg 0$, such that

$$q = \mathrm{Mult}_p(D) \geq 2k+1.$$

If p is an almost isolated singularity of $2D/q$, the linear system $|K_X+B|$ is free at p by Lemma 7. So we may assume that p is not an almost isolated singular point of $2D/q$. Denote by F the irreducible curve that has the highest multiplicity among all the components of D that go through the given point p. We may express D in the form

$$D = rF + G + R,$$

where r is the multiplicity of D along F. Since p is not an almost isolated singular point of $2D/q$, we note that $r > q/2$. This inequality implies that F is smooth at p. Denote by G the sum of the other components of D that pass through the given point p. Denote by R the sum of the other remaining components of D. We consider the $\mathbf{Q}$-divisor

$$D/r = F + N + \Delta,$$

where Δ is the fractional part of D/r and $N = [R/r]$ is an effective divisor that does not contain p. Now the vanishing theorem says that the cohomology group

$$H^1(\mathcal{O}_X(K_X+B-N-F)) = 0.$$

We conclude that the restriction map

$$H^0(\mathcal{O}_X(K_X+B-N)) \longrightarrow H^0(\mathcal{O}_X(K_X+B-N)\big|_F)$$

is surjective. We note that

$$K_X+B-N\big|_F \equiv K_F + \left(1-\tfrac{k}{r}\right)\cdot B + \Delta\big|_F.$$

We also observe that $\mathrm{Mult}_p(\Delta) = \left(\frac{1}{r}\cdot q - 1\right)$ and $F\cdot B \geq 2$. We conclude that

$$\deg\left(\left(1-\tfrac{k}{r}\right)\cdot B + \Delta\right)\big|_F > 1.$$

This implies that $K_X + B - N\big|_F$ is equivalent to a divisor of the form $K_F + C$, where $\deg(C) \geq 2$. Hence we can find a section of $\mathcal{O}_X(K_X + B - N\big|_F)$ that does not vanish at p. Since this section can be lifted to a section of $\mathcal{O}_X(K_X + B - N)$, we conclude that the linear system $|K_X + B|$ is free at p.

§3. In this last section, we would like to report on some of the recent results on adjoint linear systems. In [M], Mumford gave a proof for the Kodaira vanishing theorem by using Bogomolov's instability criterion. Recently, Fernandez del Busto has found a very elegant proof of the famous inequality of Bogomolov by using the Kawamata–Viehweg vanishing theorem and techniques similar to our proof of Reider's theorem [Fer]. It would be desirable to have a better understanding of the relations among these different results. In another development, using Damailly's theorem Siu obtained the following effective version of the Matsusaka theorem.

THEOREM 8 (Siu). *Let A be an ample divisor on a smooth projective n-fold X. Consider the intersection numbers $a = (A)^n$, $b = K_X\cdot(A)^{n-1}$ and $c = (n+2)a+b$. Assume that*

$$m > 24^n c\left((1+c)^n\right)^{n(6n^3)^n}.$$

Then the divisor mA is very ample.

Recently, Lazarsfeld, Nakamaye, and I have found an algebraic proof for the following result that replaces the hypothesis on the tangent sheaf in Demailly's theorem by a simpler assumption on $-K_X$.

THEOREM 9 (Ein, Nakamaye, and Lazarsfeld). *Let B be a nef and big divisor on a smooth projective n-fold X and let p be point in X. Assume that there is a nonnegative number m such that $-K_X + mB$ is ample. Then there exist n positive constants, $c_1, c_2, \ldots, c_n$ that can be expressed explicitly as functions of n and m alone with the following property. Suppose that the divisor B satisfies the following numerical conditions. For $k = 1, 2, \ldots, n$,*

$$B^k\cdot Y > c_k \quad \textit{for all } k\textit{-dimensional subvarieties } Y \textit{ such that } \; p \in Y.$$

Then the linear system $|K_X + B|$ is free and separates tangents at p.

REMARK. We also obtained similar results for very ampleness. These results give numerical criteria for freeness and very ampleness for ample linear systems on those varieties where $-K_X$ is nef, such as Fano and Calabi–Yau manifolds.

Using some of the ideas of Demailly and Siu, Fernandez del Busto has found the following simple form of effective Matsusaka theorem for ample line bundles on surfaces.

THEOREM 10 (Fernandez del Busto). *Let A be an ample divisor on a smooth surface X. Set $a = A \cdot A$ and $b = A \cdot K_X$. Then the linear system $|mA|$ is free for $m > (3a + b + 1)^2/(2a + 2)$.*

Let B be an ample divisor on a smooth projective n-fold and let p be a given point of X. One would like to measure the local positivity of B at p. Let $f : Y \longrightarrow X$ be the blowing up of X at p and let E be the exceptional divisor. Following Demailly, we define the Seshadri constant of B at p to be the real number

$$s(B,p) = \sup\{t \in \mathbf{Q} \mid \text{the } \mathbf{Q}\text{-divisor } f^*B - tE \text{ is ample}\}.$$

Suppose that m is a positive integer such that $m \cdot s(B,p) > n$. Then $f^*mB - nE$ is ample. Observe that

$$f^*(K_X + mB) - E = K_Y + f^*mB - nE.$$

Using the Kodaira vanishing theorem, we conclude that $|K_X + mB|$ is free at p. Similarly, one can obtain results for separating points and higher-order jets. To apply these observations, one would need to give a lower estimate on $s(B,p)$. In [EL2] we show that for X a surface we have $S(B,p) \geq 1$ for a general point p in X. Recently, together with Kuechle and Lazarsfeld, we found the following result for higher-dimensional varieties.

THEOREM 11. *Let B be an ample divisor on a smooth projective n-fold X. If p is a very general point of X, then $s(B,p) \geq \frac{1}{n}$.*

REMARK. Let r be a positive integer. If $m > n(n+r)$, Theorem 11 implies that the linear system $|K_X + mB|$ separates r jets at a general point p of X.

Together with Lazarsfeld and Masek, we have generalized Theorem 6 to threefolds with logterminal singularities. In a recent preprint, Fujita has found an improvement for Theorem 6. His result implies the following theorem [Fuj2].

THEOREM 12 (Fujita). *Let A be an ample divisor on a smooth projective threefold X. Assume that $A^3 \geq 2$. Then $|K_X + 3A|$ is free.*

In his study of varieties with nontrivial fundamental group, Kollár finds the following interesting result [Kol2].

THEOREM 13 (Kollár). *Let X be a smooth projective variety and let x be a general point in X. Assume that every irreducible positive-dimensional subvariety Z containing x has the following property. If $Z' \to Z$ is a resolution of singularities of Z, then $\pi_1(Z')$ is an infinite group. Then, for every nef and big divisor B on X, the linear system $|K_X + B| \neq \varnothing$.*

Despite all this recent progress, many fundamental problems in this area remain unsolved. We conclude with a short list of open problems.

1. Can one generalize Theorem 6 to higher dimension and prove Fujita's conjectures?
2. For threefolds, one may ask what is the optimal result and can one also prove a similar theorem for very ampleness.
3. What is the analog of Bogomolov's theorem for linear systems on a higher-dimensional variety?
4. For surfaces, what is the optimal effective Matsusaka's theorem? Can one describe the properties of the extremal examples?
5. Is the analog of the theorem in [EL2] true in higher dimension? One would like to show that for a nef and big divisor B on a smooth n-fold, $s(B,p) \geq 1$ for a general point p in X.

References

[Ben] X. Benveniste, *Sur les variétés de dimension 3 de type général dont le diviseur canonique est numériquement positif*, Math. Ann. **266** (1984), 479–497.

[Bmb] E. Bombieri, *Canonical models of surfaces of general type*, Publ. Math. IHES **42** (1973), 171–219.

[Dem] J. P. Demailly, *A numerical criterion for very ample bundles*, J. Diff. Geom. **37** (1993), 323–374.

[EL1] L. Ein and R. Lazarsfeld, *Global generation of pluricanonical and adjoint linear series on smooth projective threefolds*, J. Amer. Math. Soc. **6** (1993), 875–903.

[EL2] ———, *Seshadri constants on smooth surfaces*, preprint.

[ELM] L. Ein, R. Lazarsfeld, and V. Masek, *Global generation of linear series on terminal threefolds*, preprint.

[EV] H. Esnault and E. Viehweg, *Lectures on vanishing theorems*, DMV Seminar **20**, Birkhäuser, Basel, 1993.

[Fer] G. Fernandez del Busto, *Bogomolov's instability and Kawamata–Viehweg's vanishing*, preprint.

[Fuj1] T. Fujita, *Contribution to birational geometry of algebraic varieties: Open problems*, from the 1988 Katata Symposium of the Taniguchi Foundation.

[Fuj2] ———, *Remarks on Ein–Lazarsfeld criterion of spannedness of adjoint bundles of polarized threefold*, preprint.

[KMM] Y. Kawamata, K. Matsuda, and K. Matsuki, *Introduction to the minimal model problem*, Algebraic Geometry, Sendai, 1985 (T. Oda, ed.), Adv. Stud. Pure Math. **10**, North-Holland, Amsterdam, 1987, pp. 293–360.

[Kod] K. Kodaira, *Pluricanonical systems on algebraic surfaces of general type*, J. Math. Soc. Japan **20** (1968), 170–192.

[Kol1] J. Kollár, *Effective base point freeness*, Math. Ann. **296** (1993), 595–605.

[Kol2] ———, *Shafarevich maps and automorphic forms*, preprint.

[M] D. Mumford, *Some footnotes to the work of C. P. Ramanujam, D. P. Ramanujam, a tribute*, Springer, 1978, pp. 247–262.

[Rd] M. Reid, *Projective morphisms according to Kawamata*, unpublished manuscript (1983).

[Rdr] I. Reider, *Vector bundles of rank 2 and linear systems on algebraic surfaces*, Ann. of Math. **127** (1988), 309–316.

[S] V. Shokurov, *The non-vanishing theorem*, Math. USSR — Izv. **19** (1985), 591–607.

[Si] Y. Siu, *An effective Matsusaka big theorem*, preprint.

Lawrence Ein
Department of Mathematics, Statistics and Computer Science (M/C 249)
851 South Morgan University of Illinois at Chicago
Chicago, IL 60607-7045

E-mail address: u22425@uic.uicvm.edu

Complex Algebraic Geometry
MSRI Publications
Volume **28**, 1995

Torelli Groups and Geometry of Moduli Spaces of Curves

RICHARD M. HAIN

ABSTRACT. The Torelli group T_g is the group of isotopy classes of diffeomorphisms of a compact orientable surface of genus g that act trivially on the homology of the surface. The aim of this paper is to show how facts about the homology of the Torelli group imply interesting results about algebraic curves. We begin with an exposition of some of Dennis Johnson's work on the Torelli groups. We then show how these results imply that the Picard group of the moduli space of curves of genus $g \geq 3$ with a level-l structure is finitely generated. A classification of all "natural" normal functions over the moduli space of curves of genus $g \geq 3$ and a level l structure is obtained by combining Johnson's results with M. Saito's theory of Hodge modules. This is used to prove results that generalize the classical Franchetta Conjecture to the generic curve of genus g with n marked points and a level-l structure. Other applications are given, for example, to computing heights of cycles defined over a moduli space of curves.

1. Introduction

The Torelli group T_g is the kernel of the natural homomorphism $\Gamma_g \to \mathrm{Sp}_g(\mathbb{Z})$ from the mapping class group in genus g to the group of $2g \times 2g$ integral symplectic matrices. It accounts for the difference between the topology of $\mathcal{A}_g$, the moduli space of principally polarized abelian varieties of dimension g, and $\mathcal{M}_g$, the moduli space of smooth projective curves of genus g, and therefore should account for some of the difference between their geometries. For this reason, it is an important problem to understand its structure and its cohomology. To date, little is known about T_g apart from Dennis Johnson's few fundamental results—he has proved that T_g is finitely generated when $g \geq 3$ and has computed $H_1(T_g, \mathbb{Z})$. It is this second result that will concern us in this paper. Crudely stated, it says that there is an $\mathrm{Sp}_g(\mathbb{Z})$-equivariant isomorphism

$$H^1(T_g, \mathbb{Q}) \approx PH^3(\mathrm{Jac}\, C, \mathbb{Q}),$$

Research supported in part by grants from the National Science Foundation and the NWO.

where C is a smooth projective curve of genus g, and P denotes primitive part.

My aim in this paper is to give a detailed exposition of Johnson's homomorphism

$$PH^3(\operatorname{Jac} C, \mathbb{Q}) \to H^1(T_g, \mathbb{Q})$$

and to explain how Johnson's computation, alone and in concert with M. Saito's theory of Hodge modules [43], has some remarkable consequences for the geometry of $\mathcal{M}_g$. It implies quite directly, for example, that for each l, the Picard group of the moduli space $\mathcal{M}_g(l)$ of curves of genus $g \geq 3$ with a level-l structure is finitely generated. Combined with Saito's work, it enables one to completely write down all "natural" generically defined normal functions over $\mathcal{M}_g(l)$ when $g \geq 3$. The result is that, modulo torsion, all are half-integer multiples of the normal function of the cycle $C - C^-$. This is applied to give a new proof of the Harris–Pulte Theorem [27, 41], which relates the mixed Hodge structure on the fundamental group of a curve C to the algebraic cycle $C - C^-$ in its jacobian. Another application is to show that the cycles $C^{(a)} - i_* C^{(a)}$ in $\operatorname{Jac} C$, for $1 \leq a < g-1$, are of infinite order modulo algebraic equivalence for the general curve C. This result is due to Ceresa [10].

Understanding all normal functions over $\mathcal{M}_g(l)$ also allows us to "compute" the archimedean height pairing between any two "natural" cycles in a smooth projective variety defined over the moduli space of curves, provided they are homologically trivial over each curve, disjoint over the generic curve, and satisfy the usual dimension restrictions. The precise statement can be found in Section 14.

Our final application is to the Franchetta conjecture. The classical version of this conjecture asserts that the Picard group of the generic curve is isomorphic to $\mathbb{Z}$ and is generated by the canonical divisor. Beauville (unpublished), and later Arbarello and Cornalba [1], deduced this from Harer's computation of $H^2(\Gamma_g)$. As another application of the classification of normal functions over $\mathcal{M}_g(l)$, we prove a "Franchetta Conjecture" for the generic curve with a level-l structure. The statement is that the Picard group of the generic curve of genus g with a level-l structure is finitely generated of rank 1—the torsion subgroup is isomorphic to $(\mathbb{Z}/l\mathbb{Z})^{2g}$; mod torsion, it is generated by the canonical bundle if l is odd, and by a square root of the canonical bundle if l is even. Our proof is valid only when $g \geq 3$; it does not use the computation of $\operatorname{Pic} \mathcal{M}_g(l)$, which is not known at this time. We also compute the Picard group of the generic genus-g curve with a level-l structure and n marked points.

Our results on normal functions are inspired by those in the last section of Nori's remarkable paper [40], where he studies functions on finite covers of Zariski open subsets of the moduli space of principally polarized abelian varieties. There are analogues of our main results for $\mathcal{A}_g(l)$, the moduli space of principally polarized abelian varieties of dimension g with a level-l structure. These results are similar to Nori's, but differ. The detailed statements, as well as a discussion of the relation between the results, are in Section 15. Our results on abelian varieties are related to some results of Silverberg [45].

Sections 3 and 4 contain an exposition of the three constructions of the Johnson homomorphism that are given in [33]. Since no proof of their equivalence appears in the literature, I have given a detailed exposition, especially since the equivalence of two of these constructions is essential in one of the applications to normal functions.

In Section 13 Johnson's result is used to give an explicit description of the action of the Γ_g on the n-th roots of the canonical bundle. This is a slight refinement of a result of Sipe [46]. A consequence of this computation is that the only roots of the canonical bundle defined over Torelli space are the canonical bundle itself and all theta characteristics.

Acknowledgements. First and foremost, I would like to thank Eduard Looijenga for his hospitality and for stimulating discussions during a visit to the University of Utrecht in the spring of 1992 during which some of the work in this paper was done. I would also like to thank the University of Utrecht and the Dutch NWO for their generous support during that visit. Thanks also to Pietro Pirola and Enrico Arbarello for pointing out to me that the non-existence of sections of the universal jacobian implies the classical Franchetta conjecture. From this it was a short step to the generalizations in Section 12. Thanks also to Arnaud Beauville for useful discussions on roots of the canonical bundle and for bringing Sipe's work to my attention.

2. Mapping class groups and moduli

At this time there is no argument within algebraic geometry to compute the Picard groups of all $\mathcal{M}_g$, and one has to resort to topology to do this computation. Let S be a compact orientable surface of genus g with r boundary components and let P be an ordered set of n distinct marked points of $S - \partial S$. Denote the group of orientation-preserving diffeomorphisms of S that fix $P \cup \partial S$ pointwise by $\mathrm{Diff}^+(S, P \cup \partial S)$. Endowed with the compact-open topology, this is a topological group. The mapping class group $\Gamma^n_{g,r}$ is defined to be its group of path components:

$$\Gamma^n_{g,r} = \pi_0 \,\mathrm{Diff}^+(S, P \cup \partial S).$$

Equivalently, it is the group of isotopy classes of orientation-preserving diffeomorphisms of S that fix $P \cup \partial S$ pointwise. It is conventional to omit the decorations n and r when they are zero. So, for example, $\Gamma^n_g = \Gamma^n_{g,0}$.

The link between moduli spaces and mapping class groups is provided by Teichmüller theory. Denote the moduli space of smooth genus g curves with

n marked points by $\mathcal{M}_g^n$. Teichmüller theory provides a contractible complex manifold $\mathcal{X}_g^n$ on which Γ_g^n acts properly discontinuously—when $2g+n+2>0$, it is the space of all complete hyperbolic metrics on $S-P$ equivalent under diffeomorphisms isotopic to the identity. The quotient $\Gamma_g^n\backslash\mathcal{X}_g^n$ is analytically isomorphic to $\mathcal{M}_g^n$. It is useful to think of Γ_g^n as the orbifold fundamental group of $\mathcal{M}_g^n$. As we shall explain shortly, there is a natural isomorphism

$$H^\bullet(\mathcal{M}_g^n,\mathbb{Q})\approx H^\bullet(\Gamma_g^n,\mathbb{Q}).$$

One can compactify S by filling in the r boundary components of S by attaching disks. Denote the resulting genus-g surface by $\overline{S}$. Elements of $\Gamma_{g,r}^n$ extend canonically to $\overline{S}$ to give a homomorphism $\Gamma_{g,r}^n\to\Gamma_g$. Denote the composite

$$\Gamma_{g,r}^n\to\Gamma_g\to\operatorname{Aut}H_1(\overline{S},\mathbb{Z})$$

by ρ. Since elements of $\Gamma_{g,r}^n$ are represented by orientation-preserving diffeomorphisms, each element of $\Gamma_{g,r}^n$ preserves the intersection pairing

$$q:\Lambda^2H_1(\overline{S},\mathbb{Z})\to\mathbb{Z}.$$

Consequently, we obtain a homomorphism

$$\rho:\Gamma_{g,r}^n\to\operatorname{Aut}(H_1(\overline{S},\mathbb{Z}),q)\approx\operatorname{Sp}_g(\mathbb{Z}).$$

This homomorphism is well-known to be surjective.

Denote the moduli space of principally polarized abelian varieties of dimension g by $\mathcal{A}_g$. Since this is the quotient of the Siegel upper half plane by $\operatorname{Sp}_g(\mathbb{Z})$, it is an orbifold with orbifold fundamental group $\operatorname{Sp}_g(\mathbb{Z})$ and, as in the case of $\mathcal{M}_g^n$, there is a natural isomorphism

$$H^\bullet(\mathcal{A}_g,\mathbb{Q})\approx H^\bullet(\operatorname{Sp}_g(\mathbb{Z}),\mathbb{Q}).$$

The period map $\mathcal{M}_g^n\to\mathcal{A}_g$ is a map of orbifolds and induces ρ on fundamental groups.

The Torelli group $T_{g,r}^n$ is the kernel of the homomorphism

$$\rho:\Gamma_{g,r}^n\to\operatorname{Sp}_g(\mathbb{Z}).$$

Since ρ is surjective, we have an extension

$$1\to T_{g,r}^n\to\Gamma_{g,r}^n\to\operatorname{Sp}_g(\mathbb{Z})\to 1.$$

The Torelli group T_g encodes the differences between the topology of $\mathcal{M}_g$ and that of $\mathcal{A}_g$—between curves and abelian varieties. More formally, we have the Hochschild–Serre spectral sequence

$$H^s(\operatorname{Sp}_g(\mathbb{Z}),H^t(T_{g,r}^n))\implies H^{s+t}(\Gamma_{g,r}^n).$$

Much more (although not enough) is known about the topology of the $\mathcal{A}_g$ than about that of the $\mathcal{M}_g$. For example, the rational cohomology groups of the $\mathcal{A}_g$ stabilize as $g \to \infty$, and this stable cohomology is known by Borel's work [6]: it is a polynomial ring generated by classes $c_1, c_3, c_5, \ldots$, where c_k has degree $2k$. As with $\mathcal{A}_g$, the rational cohomology of the $\mathcal{M}_g$ is known to stabilize, as was proved by Harer [22], but the stable cohomology of the $\mathcal{M}_g$ is known only up to dimension 4; the computations are due to Harer [24, 26].

Torelli space $\mathcal{T}_g^n$ is the quotient $T_g^n \backslash \mathcal{X}_g^n$ of Teichmüller space. When $g \geq 3$, it is the moduli space of smooth projective curves C, together with n ordered distinct points and a symplectic basis of $H_1(C, \mathbb{Z})$.

The Torelli group is torsion-free. Perhaps the simplest way to see this is to note that, by standard topology, since $\mathcal{X}_g^n$ is contractible, each element of Γ_g^n of prime order must fix a point of $\mathcal{X}_g^n$. If $\phi \in \Gamma_g^n$ fixes the point corresponding to the marked curve C, there is an automorphism of C that lies in the mapping class ϕ. Since the automorphism group of a compact Riemann surface injects into $\operatorname{Aut} H^0(C, \Omega_C^1)$, and therefore into $H_1(C)$, it follows that T_g^n is torsion-free. Because of this, the Torelli space $\mathcal{T}_g^n$ is the classifying space of T_g^n.

One can view Siegel space $\mathfrak{h}_g$ as the classifying space of principally polarized abelian varieties of dimension g together with a symplectic basis of H^1. The period map therefore induces a map

$$\mathcal{T}_g^n \to \mathfrak{h}_g,$$

which is 2:1 when $g \geq 2$, and ramified along the hyperelliptic locus when $g \geq 3$.

For a finite-index subgroup L of $\mathrm{Sp}_g(\mathbb{Z})$, let $\Gamma_{g,r}^n(L)$ be the inverse image of L in $\Gamma_{g,r}^n$ under the canonical homomorphism $\Gamma_{g,r}^n \to \mathrm{Sp}_g(\mathbb{Z})$. It may be expressed as an extension

$$1 \to T_{g,r}^n \to \Gamma_{g,r}^n(L) \to L \to 1.$$

Set $\mathcal{M}_g^n(L) = \Gamma_g^n(L) \backslash \mathcal{X}_g^n$. We will call $\Gamma_{g,r}^n(L)$ the *level-L* subgroup of $\Gamma_{g,r}^n$, and we will say that points in $\mathcal{M}_g^n(L)$ are curves with a level-L structure and n marked points. The traditional moduli space of curves with a level-l structure, where $l \in \mathbb{N}^+$, is obtained by taking L to be the elements of $\mathrm{Sp}_g(\mathbb{Z})$ that are congruent to the identity mod l.

Since the Torelli groups are torsion-free, $\Gamma_{g,r}^n(L)$ is torsion-free when L is. Note, however, that by the Lefschetz fixed point formula, $\Gamma_{g,r}^n$ is torsion-free when $n + 2r > 2g + 2$, so that $\Gamma_{g,r}^n(L)$ may be torsion-free even when L is not.

PROPOSITION 2.1. *For all $g, n \geq 0$ and for each finite-index subgroup L of $\mathrm{Sp}_g(\mathbb{Z})$, there is a natural homomorphism*

$$H^\bullet(\mathcal{M}_g^n(L), \mathbb{Z}) \to H^\bullet(\Gamma_g^n(L), \mathbb{Z}),$$

which is an isomorphism when $\Gamma_g^n(L)$ is torsion-free, and is an isomorphism after tensoring with $\mathbb{Q}$ for all L.

PROOF. Set $\Gamma = \Gamma_g^n$, $\Gamma(L) = \Gamma_g^n(L)$, $\mathcal{M}(L) = \mathcal{M}_g^n(L)$ and $\mathcal{X} = \mathcal{X}_g^n$. Let $E\Gamma$ be any space on which Γ acts freely and properly discontinuously—so $E\Gamma$ is the universal covering space of some model of the classifying space of Γ. Since $\mathcal{X}$ is contractible, the quotient $E\Gamma \times_{\Gamma(L)} \mathcal{X}$ of $E\Gamma \times \mathcal{X}$ by the diagonal action of $\Gamma(L)$ is a model $B\Gamma(L)$ of the classifying space of $\Gamma(L)$. The projection $E\Gamma \times \mathcal{X} \to \mathcal{X}$ induces a map $f : B\Gamma(L) \to \mathcal{M}(L)$, which induces the map of the theorem. If $\Gamma(L)$ is torsion-free, f is a homotopy equivalence. Otherwise, choose a finite-index, torsion-free normal subgroup L' of L. Then $\Gamma(L)$ is torsion-free. Set

$$G = \Gamma(L)/\Gamma(L') \approx L/L'.$$

This is a finite group. We have the commutative diagram of Galois G-coverings

$$\begin{array}{ccc} B\Gamma(L') & \to & \mathcal{M}(L') \\ \downarrow & & \downarrow \\ B\Gamma(L) & \to & \mathcal{M}(L), \end{array}$$

where the top map is a homotopy equivalence. Thus it induces a G-equivariant isomorphism

$$H^\bullet(\mathcal{M}(L')) \to H^\bullet(B\Gamma(L')).$$

The result follows as the vertical projections induce isomorphisms

$$H^\bullet(\mathcal{M}(L), \mathbb{Q}) \xrightarrow{\sim} H^\bullet(\mathcal{M}(L'), \mathbb{Q})^G \text{ and } H^\bullet(\Gamma(L), \mathbb{Q}) \xrightarrow{\sim} H^\bullet(\Gamma(L'), \mathbb{Q})^G. \quad \square$$

The group $\Gamma_{g,r}^n(L)$ also admits a moduli interpretation when $r > 0$, even though algebraic curves have no boundary components. The idea is that a topological boundary component of a compact orientable surface should correspond to a first-order local holomorphic coordinate about a cusp of a smooth algebraic curve. Denote by $\mathcal{M}_{g,r}^n(L)$ the moduli space of smooth curves of genus g with a level-L structure and with n distinct marked points and r distinct, non-zero cotangent vectors, where the cotangent vectors do not lie over any of the marked points, and where no two of the cotangent vectors are anchored at the same point. This is a $(\mathbb{C}^*)^r$ bundle over $\mathcal{M}_g^{r+n}(L)$.

PROPOSITION 2.2. *For all finite-index subgroups L of $\mathrm{Sp}_g(\mathbb{Z})$ and for all $g, n, r \geq 0$, there is a natural homomorphism*

$$H^\bullet(\mathcal{M}_{g,r}^n(L), \mathbb{Z}) \to H^\bullet(\Gamma_{g,r}^n(L), \mathbb{Z}),$$

which is an isomorphism when $\Gamma_{g,r}^n(L)$ is torsion-free, and is an isomorphism after tensoring with $\mathbb{Q}$ for all L. $\square$

3. The Johnson homomorphism

Dennis Johnson, in a sequence of pioneering papers [30, 31, 32], began a systematic study of the Torelli groups. From the point of view of computing the cohomology of the $\mathcal{M}_g$, the most important of his results is his computation of $H_1(T_g^1)$ [32]. Let S be a compact oriented surface of genus $g \geq 3$ with a distinguished base point x_0.

THEOREM 3.1. *There is an* $\mathrm{Sp}_g(\mathbb{Z})$*-equivariant homomorphism*

$$\tau_g^1 : H_1(T_g^1, \mathbb{Z}) \to \Lambda^3 H_1(S),$$

which is an isomorphism mod 2*-torsion.*

Johnson has also computed $H_1(T_g^1, \mathbb{Z}/2\mathbb{Z})$. It is related to theta characteristics. Bert van Geemen has interesting ideas regarding its relation to the geometry of curves. A proof of Johnson's theorem is beyond our scope, but we will give three constructions of the homomorphism τ_g^1 and establish their equality.

We begin by sketching the first of these constructions. Since the Torelli group is torsion-free, there is a universal curve

$$\mathcal{C} \to \mathcal{T}_g^1$$

over Torelli space. This has a tautological section $\sigma : \mathcal{T}_g^1 \to \mathcal{C}$. There is also the jacobian

$$\mathcal{J} \to \mathcal{T}_g^1$$

of the universal curve. The universal curve can be embedded in its jacobian using the section σ—the restriction of this mapping to the fiber over the point of Torelli space corresponding to (C, x) is the Abel–Jacobi mapping

$$\nu_x : (C, x) \to (\operatorname{Jac} C,\, 0)$$

associated to (C, x). Since T_g^1 acts trivially on the first homology of the curve, the local system associated to $H_1(C)$ is framed. There is a corresponding topological trivialization of the jacobian bundle:

$$\mathcal{J} \xrightarrow{\sim} \mathcal{T}_g^1 \times \operatorname{Jac} C.$$

Let $p : \mathcal{J} \to \operatorname{Jac} C$ be the corresponding projection onto the fiber. Each element ϕ of $H_1(T_g^1, \mathbb{Z})$ can be represented by an embedded circle $\phi : S^1 \to \mathcal{T}_g^1$. Regard the universal curve $\mathcal{C}$ as subvariety of $\mathcal{J}$ via the Abel–Jacobi mapping. Then the part of the universal curve $M(\phi)$ lying over the circle ϕ is a 3-cycle in $\mathcal{J}$. The Johnson homomorphism is defined by

$$\tau_g^1(\phi) = p_*[M(\phi)] \in H_3(\operatorname{Jac} C, \mathbb{Z}) \approx \Lambda^3 H_1(C, \mathbb{Z}).$$

This definition is nice and conceptual, but is not so easy to work with. In the remainder of this section, we remake this definition without appealing to Torelli space. In the next section, we will give two more constructions of it, both due to Johnson, and prove that all three constructions agree.

Recall that the *mapping torus* of a diffeomorphism ϕ of a manifold S is the quotient $M(\phi)$ of $S \times [0,1]$ obtained by identifying $(x,1)$ with $(\phi(x),0)$:

$$M(\phi) = S \times [0,1]/\{(x,1) \sim (\phi(x),0)\}.$$

The projection $S \times [0,1] \to [0,1]$ induces a bundle projection

$$M(\phi) \to [0,1]/\{0 \sim 1\} = S^1$$

whose fiber is S and whose geometric monodromy is ϕ.

Now suppose that $\phi : (S,x_0) \to (S,x_0)$ is a diffeomorphism of S that represents an element of T_g^1. The mapping torus bundle

$$M(\phi) \to S^1$$

has a canonical section $\sigma : S^1 \to M(\phi)$, which takes $t \in S^1$ to $(x_0,t) \in M(\phi)$.

Denote $H_\bullet(S,\mathbb{R}/\mathbb{Z})$, the "jacobian" of S, by $\operatorname{Jac} S$. The next task is to embed $M(\phi)$ into $\operatorname{Jac} S$ using the section σ of base points. To this end, choose a basis $\omega_1,\dots,\omega_{2g}$ of $H^1(S,\mathbb{Z})$. This gives an identification of $\operatorname{Jac} S$ with $(\mathbb{R}/\mathbb{Z})^{2g}$. Choose closed, real-valued one-forms $w_1,\dots,w_{2g}$ representing $\omega_1,\dots,\omega_{2g}$. They have integral periods. Since ϕ acts trivially on $H^1(S)$, there are smooth functions $f_j : S \to \mathbb{R}$ such that

$$\phi^* w_j = w_j + df_j.$$

These functions are uniquely determined if we insist, as we shall, that $f_j(x_0) = 0$ for each j. Set $\vec{w} = (w_1,\dots,w_g)$ and $\vec{f} = (f_1,\dots,f_g)$. The map

$$S \times [0,1] \to \operatorname{Jac} S$$

defined by

$$(x,t) \mapsto t\vec{f}(x) + \int_{x_0}^{x} \vec{w}$$

preserves the equivalence relations of the mapping torus $M(\phi)$, and therefore induces a map

$$\nu(\phi) : (M(\phi),\sigma(S^1)) \to (\operatorname{Jac} S,\, 0).$$

Define $\tilde{\tau}(\phi)$ to be the homology class of $M(\phi)$ in $H_3(\operatorname{Jac} S,\, \mathbb{Z})$:

$$\tilde{\tau}(\phi) = \nu(\phi)_*[M(\phi)] \in \Lambda^3 H_1(S,\mathbb{Z}).$$

PROPOSITION 3.2. *Suppose ϕ, ψ are diffeomorphisms of S that act trivially on $H_1(S)$. Then:*

(a) $\tilde{\tau}(\phi)$ *is independent of the choice of representatives* $w_1, \dots, w_g$ *of the basis* $\omega_1, \dots, \omega_{2g}$ *of* $H^1(S, \mathbb{Z})$;
(b) $\tilde{\tau}(\phi)$ *is independent of the choice of basis* $\omega_1, \dots, \omega_{2g}$ *of* $H^1(S, \mathbb{Z})$;
(c) $\tilde{\tau}(\phi)$ *depends only on the isotopy class of* ϕ;
(d) $\tilde{\tau}(\phi\psi) = \tilde{\tau}(\phi) + \tilde{\tau}(\psi)$;
(e) $\tilde{\tau}(g\psi g^{-1}) = g_*\tilde{\tau}(\phi)$ *for all diffeomorphisms* g *of* S, *where* g_* *is the automorphism of* $\Lambda^3 H_1(S)$ *induced by* g.

PROOF. If $w'_1, \dots, w'_{2g}$ is another set of representatives of the ω_j, there are functions $g_j : S \to \mathbb{R}$ such that $w'_j = w_j + dg_j$ and $g_j(x_0) = 0$. For each $s \in [0,1]$, the one-form $w_j(s) = w_j + s dg_j$ is closed on S and represents ω_j. The map

$$\nu_s : M(\phi) \to \operatorname{Jac} S$$

defined using the representatives $w_j(s)$ takes (x,t) to

$$t\big(f_j(x) + s(g_j(\phi(x)) - g_j(x))\big) + s g_j(x) + \int_{x_0}^{x} w_j.$$

Since this depends continuously on s, it follows that ν_0 is homotopic to ν_1, which proves (a).

Assertion (b) follows from linear algebra. The proof of (c) is similar to that of (a).

To prove (d), observe that the quotient of $M(\phi\psi)$ obtained by identifying $(x,1)$ with $(\psi(x), \frac{1}{2})$ is the union of $M(\phi)$ and $M(\psi)$. The map $\nu(\phi\psi)$ factors through the quotient $M(\phi) \cup M(\psi)$ of $M(\phi\psi)$, and its restrictions to $M(\phi)$ and $M(\psi)$ are $\nu(\phi)$ and $\nu(\psi)$. Additivity follows.

Suppose that $g : (S, x_0) \to (S, x_0)$ is a diffeomorphism. The map $(g, \mathrm{id}) : S \times [0,1] \to S \times [0,1]$ induces a diffeomorphism

$$F(g) : M(\phi) \to M(g\phi g^{-1}).$$

To prove (e), it suffices to verify that the diagram

$$\begin{array}{ccc} M(\phi) & \xrightarrow{F(g)} & M(g\phi g^{-1}) \\ {\scriptstyle \nu(\phi)}\downarrow & & \downarrow{\scriptstyle \nu(g\phi g^{-1})} \\ \operatorname{Jac} S & \xrightarrow{g_*} & \operatorname{Jac} S \end{array}$$

commutes up to homotopy. In the proof of (a) we saw that the homotopy class of ν depends only on the basis of $H^1(S, \mathbb{Z})$ and not on the choice of de Rham representatives. Set $w'_j = g^* w_j$. Since the diagram

$$\begin{array}{ccc} H_1(S) & \xrightarrow{g_*} & H_1(S) \\ {\scriptstyle \int w'_j}\downarrow & & \downarrow{\scriptstyle \int w_j} \\ \mathbb{R} & = & \mathbb{R} \end{array}$$

commutes, it suffices to verify that the diagram

$$\begin{array}{ccc} M(\phi) & \xrightarrow{F(g)} & M(g\phi g^{-1}) \\ {\scriptstyle \nu'}\downarrow & & \downarrow{\scriptstyle \nu} \\ (\mathbb{R}/\mathbb{Z})^{2g} & \xrightarrow{\text{id}} & (\mathbb{R}/\mathbb{Z})^{2g} \end{array}$$

commutes, where ν is defined using $w_1, \dots, w_{2g}$, and ν' is defined using the representatives $w'_1, \dots, w'_{2g}$. This is easily done. □

Recall that the homology groups of T_g^1 are $\mathrm{Sp}_g(\mathbb{Z})$-modules; the action on $H_1(T_g)$ is given by

$$g : [\phi] \mapsto [\tilde{g}\phi\tilde{g}^{-1}],$$

where $g \in \mathrm{Sp}_g(\mathbb{Z})$ and $\tilde{g}$ is any element of Γ_g^1 that projects to g under the canonical homomorphism.

COROLLARY 3.3. *The map $\tilde{\tau}$ induces an $\mathrm{Sp}_g(\mathbb{Z})$-equivariant homomorphism*

$$\tau_g^1 : H_1(T_g^1, \mathbb{Z}) \to \Lambda^3 H_1(S, \mathbb{Z}). \qquad \square$$

From τ_g^1, we can construct a representation τ_g of $H_1(T_g)$. The kernel of the natural surjection $T_g^1 \to T_g$ is isomorphic to $\pi_1(S, x_0)$. The composition of the induced map $H_1(S, \mathbb{Z}) \to H_1(T_g^1, \mathbb{Z})$ with τ_g^1 is easily seen to be the canonical inclusion

$$\_ \times [S] : H_1(S, \mathbb{Z}) \hookrightarrow H_3(\mathrm{Jac}\, S,\ \mathbb{Z})$$

induced by taking Pontrjagin product with $\nu_*[S]$. We therefore have an induced $\mathrm{Sp}_g(\mathbb{Z})$-equivariant map

$$\tau_g : H_1(T_g, \mathbb{Z}) \to \Lambda^3 H_1(S, \mathbb{Z})/H_1(S, \mathbb{Z}).$$

The following result of Johnson is an immediate corollary of Theorem 3.1.

THEOREM 3.4. *The homomorphism τ_g is an isomorphism modulo 2-torsion.*

It is not difficult to bootstrap up from Johnson's basic computation to prove the following result.

THEOREM 3.5. *There is a natural $\mathrm{Sp}_g(\mathbb{Z})$-equivariant isomorphism*

$$\tau_{g,r}^n : H_1(T_{g,r}^n, \mathbb{Q}) \to H_1(S, \mathbb{Q})^{\oplus(n+r)} \oplus \Lambda^3 H_1(S, \mathbb{Q})/H_1(S, \mathbb{Q}).$$

An important consequence of Johnson's theorem is that the action of $\mathrm{Sp}_g(\mathbb{Z})$ on $H_1(T_{g,r}^n, \mathbb{Q})$ factors through a rational representation of the $\mathbb{Q}$-algebraic group Sp_g. Let $\lambda_1, \dots, \lambda_g$ be a fundamental set of dominant integral weights of Sp_g.

Denote the irreducible Sp_g-module with highest weight λ by $V(\lambda)$. The fundamental representation of Sp_g is $H_1(S)$. It is well-known (and easily verified) that

$$\Lambda^3 H_1(S) \approx V(\lambda_1) \oplus V(\lambda_3).$$

The previous result can be restated by saying that

$$H_1(T^n_{g,r}, \mathbb{Q}) \approx V(\lambda_3) \oplus V(\lambda_1)^{\oplus(n+r)}$$

as Sp_g-modules.

4. A second definition of the Johnson homomorphism

In this section we relate the definition of τ^1_g given in the previous section to Johnson's original definition, which is defined using the action of T^1_g on the lower central series of $\pi_1(S, x_0)$. It is better suited to computations. In order to relate this definition to the one given in the previous section, we need to study the cohomology ring of the mapping torus associated to an element of the Torelli group.

Suppose that the diffeomorphism $\phi : (S, x_0) \to (S, x_0)$ represents an element of T^1_g. As explained in the previous section, the associated mapping torus $M = M(\phi)$ fibers over S^1 and has a canonical section σ. These data guarantee that there is a canonical decomposition of the cohomology of M.

Since ϕ acts trivially on the homology of S, the E_2-term of the Leray–Serre spectral sequence of the fibration $\pi : M \to S^1$ satisfies

$$E_2^{r,s} = H^r(S^1) \otimes H^s(S).$$

This spectral sequence degenerates for trivial reasons. Consequently, there is a short exact sequence

$$0 \to H^1(S^1, \mathbb{Z}) \xrightarrow{\pi^*} H^1(M, \mathbb{Z}) \xrightarrow{i^*} H^1(S, \mathbb{Z}) \to 0,$$

where π is the projection to S^1 and $i : S \hookrightarrow M$ is the inclusion of the fiber over the base point $t = 0$ of S^1. The section σ induces a splitting of this sequence. Denote π^* of the positive generator of $H^1(S^1, \mathbb{Z})$ by θ. Then we have the decomposition

$$H^1(M, \mathbb{Z}) = H^1(S, \mathbb{Z}) \oplus \mathbb{Z}\theta. \tag{1}$$

From the spectral sequence, it follows that we have an exact sequence

$$0 \to \theta \wedge H^1(S, \mathbb{Z}) \to H^2(M, \mathbb{Z}) \xrightarrow{i^*} H^2(S, \mathbb{Z}) \to 0.$$

Denote the Poincaré dual of a homology class u in M by $\mathrm{PD}(u)$. Since

$$\int_S \mathrm{PD}(\sigma) = \sigma \cdot S = 1,$$

it follows that the previous sequence can be split by taking the positive generator of $H^2(S,\mathbb{Z})$ to $\mathrm{PD}(\sigma)$. We therefore have a canonical splitting

$$H^2(M,\mathbb{Z}) = \mathbb{Z}\mathrm{PD}(\sigma) \oplus \theta \wedge H^1(S,\mathbb{Z}). \tag{2}$$

The cup product pairing

$$c : H^1(M) \otimes H^2(M) \to H^3(M) \approx \mathbb{Z}$$

induces pairings between the summands of the decompositions (1) and (2).

PROPOSITION 4.1. *The cup product c satisfies:*
(a) $c(\theta \otimes \mathrm{PD}(\sigma)) = 1$;
(b) *the restriction of c to $H^1(S) \otimes \mathrm{PD}(\sigma)$ vanishes;*
(c) *the restriction of c to $\theta \otimes \big(\theta \wedge H^1(S)\big)$ vanishes;*
(d) *the restriction of c to $H^1(S) \otimes \big(\theta \wedge H^1(S)\big)$ takes $u \otimes (\theta \wedge v)$ to $-\int_S u \wedge v$.*

PROOF. Since θ is the Poincaré dual of the fiber S, we have

$$\int_M \theta \wedge \mathrm{PD}(\sigma) = \int_M \mathrm{PD}(S) \wedge \mathrm{PD}(\sigma) = S \cdot \sigma = 1.$$

In the decomposition (1), $H^1(S)$ is identified with the kernel of $\sigma^* : H^1(M) \to H^1(S^1)$; that is, with those $u \in H^1(M)$ such that

$$\int_\sigma u = 0.$$

The second assertion now follows as

$$\int_M u \wedge \mathrm{PD}(\sigma) = \int_\sigma u$$

for all $u \in H^1(M)$.

The third and fourth assertions are easily verified. □

To complete our understanding of the cohomology ring of M, we consider the cup product

$$\Lambda^2 H^1(M) \to H^2(M).$$

Since $\theta \wedge \theta = 0$, there is only one interesting part of this mapping, namely, the component

$$\Lambda^2 H^1(S) \to \mathbb{Z}\mathrm{PD}(\sigma) \oplus \theta \wedge H^1(S).$$

There is a unique function

$$f_\phi : \Lambda^2 H^1(S,\mathbb{Z}) \to H^1(S,\mathbb{Z})$$

such that

$$u \wedge v \mapsto \left(\int_S u \wedge v, -\theta \wedge f_\phi(u \wedge v)\right) \in H^2(M,\mathbb{Z})$$

with respect to the decomposition (2). We can view f_ϕ as an element of

$$H_1(S,\mathbb{Z}) \otimes \Lambda^2 H^1(S,\mathbb{Z}).$$

Using Poincaré duality on the last two factors, we can regard f_ϕ as an element $F(\phi)$ of

$$H_1(S,\mathbb{Z}) \otimes \Lambda^2 H_1(S,\mathbb{Z}).$$

There is a canonical embedding of $\Lambda^3 H_1(S,\mathbb{Z})$ into this group. It is defined by

$$a \wedge b \wedge c \mapsto a \otimes (b \wedge c) + b \otimes (c \wedge a) + c \otimes (a \wedge b).$$

THEOREM 4.2. *The invariant $F(\phi)$ of the cohomology ring of $M(\phi)$ is the image of $\tilde{\tau}(\phi)$ under the canonical embedding*

$$\Lambda^3 H_1(S,\mathbb{Z}) \hookrightarrow H_1(S,\mathbb{Z}) \otimes \Lambda^2 H_1(S,\mathbb{Z}).$$

PROOF. The dual of $\tau_g^1(\phi)$ is the map

$$\Lambda^3 H^1(S) \to \mathbb{Z}$$

defined by

$$u \wedge v \wedge w \mapsto \int_{M(\phi)} u \wedge v \wedge w.$$

Here we have identified $H^1(S)$ with $H^1(\operatorname{Jac} S)$ using the canonical isomorphism

$$\nu^* : H^1(\operatorname{Jac} S) \xrightarrow{\sim} H^1(S).$$

The map $\nu(\phi) : M \to \operatorname{Jac} S$ collapses σ to the point 0. It follows that the image of

$$\nu(\phi)^* : H^1(\operatorname{Jac} S) \to H^1(M)$$

lies in the subspace we are identifying with $H^1(S)$ in the decomposition (1) of page 107. Since the restriction of $\nu(\phi)$ to the fiber over the base point $t = 0$ of S^1 is the isomorphism ν^*, it follows that the diagram

$$\begin{array}{ccc} H^1(\operatorname{Jac} S) & \xrightarrow{\nu(\phi)^*} & H^1(M) \\ {\scriptstyle \nu^*}\downarrow & & \| \\ H^1(S) & \xrightarrow{i} & H^1(M) \end{array}$$

commutes, where i is the inclusion given by the splitting (1). That is, all the identifications we have made with $H^1(S)$ are compatible.

We will compute the dual of $\tau_g^1(\phi)$ using $F(\phi)$, which we regard as a homomorphism

$$F(\phi) : H^1(S) \otimes \Lambda^2 H^1(S) \to \mathbb{Z}.$$

It follows from Proposition 4.1 that this map takes $u \otimes (v \wedge w)$ to

$$\int_S u \wedge f_\phi(v \wedge w).$$

The assertion that $F(\phi)$ lies in $\Lambda^3 H_1(S)$ is equivalent to the assertion that

$$F(\phi)(u \otimes (v \wedge w)) = F(\phi)(v \otimes (w \wedge u)) = F(\phi)(w \otimes (u \wedge v)),$$

which is easily verified using Proposition 4.1. The equality of $F(\phi)$ and $\tau_g^1(\phi)$ follows as

$$\tau_g^1(\phi)(u \wedge v \wedge w) = \int_M u \wedge v \wedge w = -\int_M u \wedge \theta \wedge f_\phi(v \wedge w) = F(\phi)(u \otimes (v \wedge w)).$$

□

We are now ready to give Johnson's original definition of τ_g^1. Denote the lower central series of a group π by

$$\pi = \pi^{(1)} \supseteq \pi^{(2)} \supseteq \pi^{(3)} \supseteq \cdots$$

We regard the cup product

$$\Lambda^2 H^1(S, \mathbb{Z}) \to H^2(S, \mathbb{Z}) \approx \mathbb{Z}$$

as an element q of $\Lambda^2 H_1(S, \mathbb{Z})$.

Proposition 4.3. *The commutator mapping*

$$[\ ,\ ] : \pi_1(S, x_0) \times \pi_1(S, x_0) \to \pi_1(S, x_0)$$

induces an isomorphism $\Lambda^2 H_1(S, \mathbb{Z})/q \to \pi_1(S, x_0)^{(2)}/\pi_1(S, x_0)^{(3)}$.

Proof. This follows directly from the standard fact (see [44] or [36]) that if F is a free group, the commutator induces an isomorphism

$$\Lambda^2 H_1(F) \xrightarrow{\sim} F^{(2)}/F^{(3)}$$

and from the standard presentation of $\pi_1(S, x_0)$. □

An element ϕ of T_g^1 induces an automorphism of $\pi_1(S, x_0)$. Since it acts trivially on $H_1(S)$,

$$\phi(\gamma)\gamma^{-1} \in \pi_1(S, x_0)^{(2)}$$

for all $\gamma \in \pi_1(S, x_0)$. From Proposition 4.3, it follows that ϕ induces a well defined map

$$\hat{\tau}(\phi) : H_1(S, \mathbb{Z}) \to \Lambda^2 H_1(S, \mathbb{Z})/q$$

Using Poincaré duality, we may view this as an element $L(\phi)$ of

$$H_1(S, \mathbb{Z}) \otimes \left(\Lambda^2 H_1(S, \mathbb{Z})/q\right).$$

Theorem 4.4. *The image of* $F(\phi)$ *in* $H_1(S, \mathbb{Z}) \otimes \left(\Lambda^2 H_1(S, \mathbb{Z})/q\right)$ *is* $L(\phi)$.

PROOF. Since $H_1(S,\mathbb{Z}) \otimes (\Lambda^2 H_1(S,\mathbb{Z})/q)$ is torsion-free, it suffices to show that the image of $F(\phi)$ in

$$H_1(S,\mathbb{Q}) \otimes (\Lambda^2 H_1(S,\mathbb{Q})/q)$$

is $L(\phi)$. For the rest of this proof, all (co)homology groups have $\mathbb{Q}$ coefficients.

For all groups π with finite-dimensional $H_1(\ ,\mathbb{Q})$, the sequence

$$(3) \qquad 0 \to H^1(\pi) \xrightarrow{h^*} \left(\pi/\pi^{(3)}\right)^* \xrightarrow{[\ ,\ ]^*} \Lambda^2 H^1(\pi) \xrightarrow{\wedge} H^2(\pi)$$

of $\mathbb{Q}$ vector spaces is exact. Here $(\ )^*$ denotes the dual vector space, h^* the dual Hurewicz homomorphism, and $[\ ,\ ]$ the map induced by the commutator. This can be proved using results in either [11, §2.1] or [48, §8].

We apply this sequence to the fundamental group of the mapping torus. Choose $m_0 = (x_0, 0)$ as the base point of M. Since M fibers over the circle with fiber S, we have an exact sequence

$$1 \to \pi_1(S,x_0) \to \pi_1(M,m_0) \to \mathbb{Z} \to 0.$$

The section σ induces a splitting $\mathbb{Z} \to \pi_1(M,m_0)$.

Denote the image of 1 by σ. Observe that if $\gamma \in \pi_1(S,x_0)$, then

$$\sigma\gamma\sigma^{-1} = \phi(\gamma).$$

It follows that the inclusion $\pi_1(S,x_0) \hookrightarrow \pi_1(M,m_0)$ induces isomorphisms

$$\pi_1(S,x_0)^{(k)} \approx \pi_1(M,m_0)^{(k)}$$

for all $k > 1$ and, as above, that σ induces an isomorphism

$$H_1(M) = H_1(S) \oplus \mathbb{Q}\Sigma,$$

where Σ denotes the homology class of σ. It also follows that for all $a \in H_1(S)$

$$\tilde{\tau}(\phi)(a) = [\Sigma, a] \in \pi_1(M)^{(2)}/\pi_1(M)^{(3)} \approx \pi_1(S)^{(2)}/\pi_1(S)^{(3)}.$$

Using Proposition 4.1 and the exact sequence (3), we see that for all $u \in H_1(S)$, the image of $f_\phi^*(u)$ in

$$\Lambda^2 H_1(S)/q$$

is $[\Sigma, u]$, which is $\hat{\tau}(\phi)(u)$ as we have seen. The result follows. □

The composite of the inclusion

$$\Lambda^3 H_1(S,\mathbb{Z}) \hookrightarrow H_1(S,\mathbb{Z}) \otimes \Lambda^2 H_1(S,\mathbb{Z})$$

with the quotient mapping

$$H_1(S,\mathbb{Z}) \otimes \Lambda^2 H_1(S,\mathbb{Z}) \to H_1(S,\mathbb{Z}) \otimes (\Lambda^2 H_1(S,\mathbb{Z})/q)$$

is injective. One way to see this is to tensor with $\mathbb{Q}$ and note that both of these maps are maps of Sp_g-modules. One can then use the fact that $\Lambda^3 H_1(S)$ is the sum of the first and third fundamental representations of Sp_g to check the result. The following result is therefore a restatement of Theorem 4.2.

COROLLARY 4.5. *$L(\phi)$ lies in the image of the canonical injection*

$$\Lambda^3 H_1(S,\mathbb{Z}) \hookrightarrow H_1(S,\mathbb{Z}) \otimes \left(\Lambda^2 H_1(S,\mathbb{Z})/q\right)$$

and the corresponding point of $\Lambda^3 H_1(S)$ is $\tau_g^1(\phi)$.

In his fundamental papers, Johnson defines $\tau_g^1(\phi)$ to be $L(\phi)$. The other two definitions we have given were stated in [33].

5. Picard groups

In [39], Mumford showed that

$$c_1 : \mathrm{Pic}\,\mathcal{M}_g \otimes \mathbb{Q} \to H^2(\mathcal{M}_g, \mathbb{Q})$$

is an isomorphism. Using Johnson's computation of $H_1(T_g,\mathbb{Q})$ and the well-known Theorem 5.3, we will prove the analogous statement for all $\mathcal{M}_{g,r}^n(L)$ when $g \geq 3$. The novelty lies in the variation of the level, and not in the variation of the decorations r and n. The first, and principal, step is to establish the vanishing of the $H^1(\mathcal{M}_{g,r}^n(L))$.

PROPOSITION 5.1. *Suppose that L is a finite-index subgroup of $\mathrm{Sp}_g(\mathbb{Z})$. If $g \geq 3$, then $H^1(\mathcal{M}_{g,r}^n(L), \mathbb{Z}) = 0$.*

Since $H^1(\ ,\mathbb{Z})$ is always torsion-free, it suffices to prove that $H^1(\mathcal{M}_g(L),\mathbb{Q})$ vanishes. We will prove a stronger result.

PROPOSITION 5.2. *Suppose that L is a finite-index subgroup of $\mathrm{Sp}_g(\mathbb{Z})$ and that $g \geq 3$. If $V(\lambda)$ is an irreducible representation of Sp_g with highest weight λ, then*

$$H^1(\Gamma_{g,r}^n(L), V(\lambda)) = \begin{cases} \mathbb{Q}^{r+n} & \text{if } \lambda = \lambda_1; \\ \mathbb{Q} & \text{if } \lambda = \lambda_3; \\ 0 & \text{otherwise.} \end{cases}$$

Consequently, $H^1(\mathcal{M}_{g,r}^n(L),\mathbb{Z})$ vanishes for all r and n when $g \geq 3$.

PROOF. It follows from the Hochschild–Serre spectral sequence

$$H^r(L, H^s(T_{g,r}^n \otimes V(\lambda))) \implies H^{r+s}(\Gamma_{g,r}^n(L), V(\lambda))$$

that there is an exact sequence

$$0 \to H^1(L, V(\lambda)) \to H^1(\Gamma_{g,r}^n(L), V(\lambda)) \to H^0(L, H^1(T_{g,r}^n \otimes V(\lambda))) \xrightarrow{d_2} H^2(L, V(\lambda)).$$

By a result of Ragunathan [42], the first term vanishes when $g \geq 2$. By Theorem 3.5, the third term vanishes except when λ is either λ_1 or λ_3. This proves the result except when λ is either λ_1 or λ_3. In these exceptional cases, the third term has rank $r+n$ or 1, respectively. To complete the proof, we need to show that the differential d_2 is zero.

There are several ways to do this. Perhaps the most straightforward is to use the result, due to Borel [7], that asserts that the last group vanishes when $g \geq 8$. This establishes the result when $g \geq 8$. When $r \geq 1$, the vanishing of d_r for all $g \geq 3$ follows from the fact that the diagram

$$\begin{array}{ccc} H^0(L, H^1(T^n_{g,r} \otimes V(\lambda))) & \xrightarrow{d_2} & H^2(L, V(\lambda)) \\ \uparrow & & \uparrow \\ H^0(L, H^1(T^n_{g+8,r} \otimes V(\lambda))) & \xrightarrow{d_2} & H^2(L_{g+8}, V(\lambda)) \end{array}$$

commutes. Here L_{g+8} is any finite-index subgroup of $\mathrm{Sp}_{g+8}(\mathbb{Z})$ such that

$$L_{g+8} \cap \mathrm{Sp}_g(\mathbb{Z}) \subseteq L$$

and the vertical maps are induced by the "stabilization map"

$$\Gamma^n_{g,r}(L) \to \Gamma^n_{g+8,\,r}(L_{g+8}).$$

When $r = 0$ and $\lambda = \lambda_1$, there is nothing to prove. This leaves only the case $r = 0$ and $\lambda = \lambda_3$, which follows from the fact that the diagram

$$\begin{array}{ccc} H^0(L, H^1(T^n_g \otimes V(\lambda))) & \xrightarrow{d_2} & H^2(L, V(\lambda)) \\ \uparrow & & \uparrow \\ H^0(L, H^1(T^n_{g,1} \otimes V(\lambda))) & \xrightarrow{d_2} & H^2(L, V(\lambda)) \end{array},$$

which arises from the homomorphism $\Gamma^n_{g,1} \to \Gamma^n_g$, commutes. □

Denote the category of $\mathbb{Z}$ mixed Hodge structures by $\mathcal{H}$. Denote the group of "integral $(0,0)$ elements" $\mathrm{Hom}_{\mathcal{H}}(\mathbb{Z}, H)$ of a mixed Hodge structure H by ΓH.

Suppose that X is a smooth variety. Since $H^1(X, \mathbb{Z})$ is torsion-free, we can define

$$W_1 H^1(X, \mathbb{Z}) = W_1 H^1(X, \mathbb{Q}) \cap H^1(X, \mathbb{Z}).$$

This is a polarized, torsion-free Hodge structure of weight 1. Set

$$JH^1(X) = \frac{W_1 H^1(X, \mathbb{C})}{W_1 H^1(X, \mathbb{Z}) + F^1 W_1 H^1(X, \mathbb{C})}.$$

This is a polarized Abelian variety.

THEOREM 5.3. *If X is a smooth variety, there is a natural exact sequence*

$$0 \to JH^1(X) \to \mathrm{Pic}\, X \to \Gamma H^2(X, \mathbb{Z}(1)) \to 0.$$

Alternatively, this theorem may be stated as saying that the cycle map

$$\operatorname{Pic} X \to H^2_{\mathcal{H}}(X, \mathbb{Z}(1))$$

is an isomorphism, where $H^\bullet_{\mathcal{H}}$ denotes Beilinson's absolute Hodge cohomology, the refined version of Deligne cohomology defined in [4].

PROOF. Choose a smooth completion $\overline{X}$ of X for which $\overline{X} - X$ is a normal crossings divisor D in $\overline{X}$ with smooth components. Denote the dimension of X by d. From the usual exponential sequence, we have a short exact sequence

$$0 \to JH^1(\overline{X}) \to \operatorname{Pic} \overline{X} \to \Gamma H^2(X, \mathbb{Z}) \to 0.$$

From [12, (1.8)], we have an exact sequence

$$CH^0(D) \to \operatorname{Pic} \overline{X} \to \operatorname{Pic} X \to 0.$$

The Gysin sequence

$$0 \to H^1(\overline{X}) \to H^1(X) \to H_{2d-2}(D)(-2d) \to H^2(\overline{X}) \to \\ H^2(X) \to H_{2d-3}(D)(-2d)$$

is an exact sequence of $\mathbb{Z}$ Hodge structures. Since $H_{2d-2}(D)(-2d)$ is torsion-free and of weight 2, it follows that

$$W_1 H^1(X, \mathbb{Z}) = H^1(\overline{X}, \mathbb{Z}),$$

and therefore that $JH^1(X) = JH^1(\overline{X})$. Next, since each component D_i of D is smooth, it follows that

$$H_{2d-3}(D)(-2d) = \bigoplus_i H^1(D_i, \mathbb{Z})(-1),$$

and is therefore torsion-free and of weight 3. It follows that the sequence

$$H_{2d-2}(D)(-2d) \to \Gamma H^2(\overline{X}) \to \Gamma H^2(X) \to 0$$

is exact. Since the cycle map

$$CH^0(D) \to H_{2d-2}(D)$$

is an isomorphism [12, (1.5)], the result follows. □

It is now an easy matter to show that the Picard groups of the $\mathcal{M}^n_{g,r}(L)$ are finitely generated.

THEOREM 5.4. *Suppose that L is a finite-index subgroup of $\mathrm{Sp}_g(\mathbb{Z})$. If $g \geq 3$, then for all r, n, the Chern class map*

$$c_1 : \operatorname{Pic} \mathcal{M}^n_{g,r}(L) \to \Gamma H^2(\mathcal{M}^n_{g,r}(L), \mathbb{Z})$$

is an isomorphism when $\Gamma^n_{g,r}(L)$ is torsion-free, and is an isomorphism after tensoring with $\mathbb{Q}$ in general.

PROOF. The case when $\Gamma^n_{g,r}(L)$ is torsion-free follows directly from Proposition 5.1 and Theorem 5.3. To prove the assertion in general, choose a finite-index normal subgroup L' of L such that $\Gamma^n_{g,r}(L')$ is torsion-free. Let

$$G = \Gamma^n_{g,r}(L)/\Gamma^n_{g,r}(L') \approx L/L'.$$

Then it follows from the Teichmüller description of moduli spaces that the projection

$$\pi : \mathcal{M}^n_{g,r}(L') \to \mathcal{M}^n_{g,r}(L)$$

is a Galois covering with Galois group G. It follows from the first case that

$$c_1 : \operatorname{Pic} \mathcal{M}^n_{g,r}(L') \to \Gamma H^2(\mathcal{M}^n_{g,r}(L'), \mathbb{Z})$$

is a G-equivariant isomorphism. The result now follows as the projection π induces isomorphisms

$$\operatorname{Pic} \mathcal{M}^n_{g,r}(L) \otimes \mathbb{Q} \approx H^0(G, \operatorname{Pic} \mathcal{M}^n_{g,r}(L') \otimes \mathbb{Q})$$

and

$$\Gamma H^2(\mathcal{M}^n_{g,r}(L), \mathbb{Q}) \approx \Gamma H^0(G, H^2(\mathcal{M}^n_{g,r}(L'), \mathbb{Q})).$$

□

If we knew that $H^2(T_g, \mathbb{Q})$ were finite-dimensional and a rational representation of Sp_g, we would know from Borel's work [6] that $H^2(\mathcal{M}^n_{g,r}(L), \mathbb{Q})$ would be independent of the level L, once g is sufficiently large; $g \geq 8$ should do it [7]. It would then follow, for sufficiently large g, that the Picard number of $\mathcal{M}^n_{g,r}(L)$ is $n+r+1$. At present it is not even known whether $H^2(T_g, \mathbb{Q})$ is finite-dimensional. The computation of this group, and the related problem of finding a presentation of T_g, appear to be deep and difficult. It should be mentioned that the only evidence for the belief that the Picard number of each $\mathcal{M}_g(L)$ is one comes from Harer's computation [25] of the Picard numbers of the moduli spaces of curves with a distinguished theta characteristic.

6. Normal functions

In this section, we define abstract normal functions that generalize the normal functions of Poincaré and Griffiths. We begin by reviewing how a family of homologically trivial algebraic cycles in a family of smooth projective varieties gives rise to a normal function.

Suppose that X is a smooth variety. A homologically trivial algebraic d-cycle in X canonically determines an element of

$$\operatorname{Ext}^1_{\mathcal{H}}(\mathbb{Z}, H_{2d+1}(X, \mathbb{Z}(-d))).$$

This extension is obtained by pulling back the exact sequence

$$0 \to H_{2d+1}(X, \mathbb{Z}(-d)) \to H_{2d+1}(X, Z, \mathbb{Z}(-d)) \to H_{2d}(Z, \mathbb{Z}(-d)) \to \cdots$$

of mixed Hodge structures along the inclusion

$$\mathbb{Z} \to H_{2d}(|Z|, \mathbb{Z}(-d))$$

that takes 1 to the class of Z.

When H is a mixed Hodge structure all of whose weights are non-positive, there is a natural isomorphism

$$JH \approx \operatorname{Ext}^1_{\mathcal{H}}(\mathbb{Z}, H),$$

where

$$JH = \frac{H_{\mathbb{C}}}{F^0 H_{\mathbb{C}} + H_{\mathbb{Z}}}.$$

(This is well-known; see [9], for example. Our conventions will be taken from [18, (2.2)].)

When X is projective, Poincaré duality provides an isomorphism of the complex torus $JH_{2d+1}(X, \mathbb{Z}(-d))$ with the Griffiths intermediate jacobian

$$\operatorname{Hom}_{\mathbb{C}}(F^d H^{d+1}(X), \mathbb{C})/H_{2d+1}(X, \mathbb{Z}).$$

The point in $JH_{2d+1}(X, Z(d))$ corresponding to the cycle Z under this isomorphism is $\int_\Gamma$, where Γ is a real $2d+1$ chain that satisfies $\partial\Gamma = Z$.

Now suppose that $\mathcal{X} \to T$ is a family of smooth projective varieties over a smooth base T. Suppose that $\mathcal{Z}$ is an algebraic cycle in $\mathcal{X}$, which is proper over T of relative dimension d. Denote the fibers of $\mathcal{X}$ and $\mathcal{Z}$ over $t \in T$ by X_t and Z_t.

The set of $H_{2d+1}(X_t, \mathbb{Z}(-d))$ form a variation of Hodge structure $\mathbb{V}$ over T of weight -1. We can form the relative intermediate jacobian

$$\mathcal{J}_d \to T,$$

which has fiber $JH_{2d+1}(X_t, \mathbb{Z}(-d))$ over $t \in T$. The family of cycles $\mathcal{Z}$ defines a section of this bundle. Such a section is what Griffiths calls *the normal function of the cycle* $\mathcal{Z}$ [14]. Griffiths' normal functions generalize those of Poincaré.

We will generalize this notion further. Before we do, note that the elements of $\operatorname{Ext}^1_{\mathcal{H}}(\mathbb{Z}, H_{2d+1}(X_t, \mathbb{Z}(-d)))$ defined by the cycles Z_t fit together to form a variation of mixed Hodge structure over T. It follows from the main result of [15] that this variation is good in the sense of [47] along each curve in T, and is therefore good in the sense of Saito [43].

Suppose that T is a smooth variety and that $\mathbb{V} \to T$ is a variation of Hodge structure over T of negative weight. Denote by $J\mathcal{V}$ the bundle over T whose fiber over $t \in T$ is

$$JV_t \approx \operatorname{Ext}^1_{\mathcal{H}}(\mathbb{Z}, V_t).$$

DEFINITION 6.1. A holomorphic section $s : T \to J\mathcal{V}$ of $J\mathcal{V} \to T$ is a *normal function* if it defines an extension

$$0 \to \mathbb{V} \to \mathbb{E} \to \mathbb{Z}_T \to 0$$

in the category $\mathcal{H}(T)$ of good variations of mixed Hodge structure over T.

REMARK 6.2. We know from the preceding discussion that families of homologically trivial cycles in a family $\mathcal{X} \to T$ define normal functions in this sense.

The asymptotic properties of good variations of mixed Hodge structure guarantee that these normal functions have nice properties.

LEMMA 6.3 (RIGIDITY). *If $\mathbb{V} \to T$ and $\mathbb{V}' \to T$ are two good variations of mixed Hodge structure over T with the same fiber V_{t_0} (viewed as a mixed Hodge structure) over some point t_0 of T and with the same monodromy representations*

$$\pi_1(T, t_0) \to \operatorname{Aut} V_{t_0},$$

then $\mathbb{V}_1$ and $\mathbb{V}_2$ are isomorphic as variations.

PROOF. The proof is a standard application of the theorem of the fixed part. The local system $\operatorname{Hom}_{\mathbb{Z}}(\mathbb{V}, \mathbb{V}')$ underlies a good variation of mixed Hodge structure. From Saito's work [43], we know that each cohomology group of a variety with coefficients in a good variation of mixed Hodge structure has a natural mixed Hodge structure. So, in particular,

$$H^0(T, \operatorname{Hom}_{\mathbb{Z}}(\mathbb{V}, \mathbb{V}'))$$

has a mixed Hodge structure, and the restriction map

$$H^0(T, \operatorname{Hom}_{\mathbb{Z}}(\mathbb{V}, \mathbb{V}')) \to \operatorname{Hom}_{\mathbb{Z}}(V_{t_0}, V'_{t_0})$$

is a morphism. The result now follows since there are natural isomorphisms

$$H^0(T, \operatorname{Hom}_{\mathbb{Z}}(\mathbb{V}, \mathbb{V}')) \approx \operatorname{Hom}_{\mathbb{Z}\pi_1(T,t_0)}(V_{t_0}, V'_{t_0})$$

and

$$\Gamma H^0(T, \mathrm{Hom}_{\mathbb{Z}}(\mathbb{V}, \mathbb{V}')) \approx \mathrm{Hom}_{\mathcal{H}(T)}(\mathbb{V}, \mathbb{V}'),$$

where $\mathcal{H}(T)$ denotes the category of good variations of mixed Hodge structure over T. □

COROLLARY 6.4. *Two normal functions $s_1, s_2 : T \to J\mathcal{V}$ are equal if and only if there is a point $t_0 \in T$ such that $s_1(t_0) = s_2(t_0)$ and such that the two induced homomorphisms*

$$(s_j)_* : \pi_1(T, t_0) \to \pi_1(J\mathcal{V}, s_1(t_0))$$

are equal. □

7. Extending normal functions

The strong asymptotic properties of variations of mixed Hodge structure imply that almost all normal functions extend across subvarieties where the original variation of Hodge structure is non-singular. Suppose that X is a smooth variety and that $\mathbb{V}$ is a variation of Hodge structure over X of negative weight. Denote the associated intermediate jacobian bundle by $\mathcal{J} \to X$.

THEOREM 7.1. *Suppose that U is a Zariski open subset of X and $s : U \to \mathcal{J}|_U$ is a normal function defined on U. If the weight of $\mathbb{V}$ is not -2, then s extends to a normal function $\tilde{s} : X \to \mathcal{J}$.*

PROOF. Write $U = X - Z$. By Hartog's Theorem, it suffices to show that s extends to a normal function on the complement of the union of the singular locus of Z and the union of the components of Z of codimension ≥ 2 in X. That is, we may assume that Z is a smooth divisor.

The problem of extending s is local. By taking a transverse slice, we can reduce to the case where X is the unit disk Δ and Z is the origin. In this case, we have a variation of Hodge structure over Δ. The normal function $s : \Delta^* \to \mathcal{J}$ corresponds to a good variation of mixed Hodge structure $\mathbb{E}$ over the punctured disk Δ^*, which is an extension

$$0 \to \mathbb{V}|_{\Delta^*} \to \mathbb{E} \to \mathbb{Z}_{\Delta^*} .$$

To prove that the normal function extends, it suffices to show that the monodromy of $\mathbb{E}$ is trivial, for then the local system $\mathbb{E}$ extends uniquely as a flat bundle to Δ and the Hodge filtration extends across the origin as $\mathbb{E}$ is a good variation.

Since $\mathbb{V}$ is defined on the whole disk, it has trivial monodromy. It follows that the local monodromy operator T of $\mathbb{E}$ satisfies

$$(T - I)^2 = 0$$

and that the local monodromy logarithm N is $T-I$. Since E is a good variation, it has a relative weight filtration $M_\bullet$ [47], which is defined over $\mathbb{Q}$ and satisfies $NM_l \subseteq M_{l-2}$. From the defining properties of $M_\bullet$ ([47, (2.5)]), we have

$$M_0 = \mathbb{E},\ M_m = \mathbb{V},\ \text{and } M_{m-1} = 0,$$

where m is the weight of $\mathbb{V}$.

In the case $m = -1$, the proof that $N = 0$ is simpler. Since this case is the most important (as it is the one that applies to normal functions of cycles), we prove it first. The condition $m = -1$ implies that $M_{-2} = 0$. Since $NM_0 \subseteq M_{-2}$, it follows that $N = 0$ and consequently, that the normal function extends.

In general, we use the defining property [47, (3.13.iii)] of good variations of mixed Hodge structure, which says that

$$(E_t, M_\bullet, F^\bullet_{\text{lim}})$$

is a mixed Hodge structure and N is a morphism of mixed Hodge structures of type $(-1,-1)$, where $F^\bullet_{\text{lim}}$ denotes the limit Hodge filtration. In this case, N induces a morphism

$$\mathbb{Z} \approx \mathrm{Gr}^M_0 \to \mathrm{Gr}^M_{-2},$$

which is zero if $m \neq -2$. Since N is a morphism of mixed Hodge structures, the vanishing of this map implies the vanishing of N. □

When $m = -2$, there are normal functions that don't extend. For example, if we take $\mathbb{V} = \mathbb{Z}(1)$, the bundle of intermediate jacobians is the bundle $X \times \mathbb{C}^*$ and the normal functions are precisely the invertible regular functions $f : X \to \mathbb{C}$. For details see, for example, [20, (9.3)].

8. Normal functions over $\mathcal{M}^n_{g,r}(L)$

Throughout this section, we will assume that $g \geq 3$ and L is a finite-index subgroup of $\mathrm{Sp}_g(\mathbb{Z})$ such that $\Gamma^n_{g,r}(L)$ is torsion-free. With this condition on L, $\mathcal{M}^n_{g,r}(L)$ is smooth. Each irreducible representation of Sp_g defines a polarized $\mathbb{Q}$ variation of Hodge structure over $\mathcal{M}^n_{g,r}(L)$, which is unique up to Tate twist: see Proposition 9.1. It follows that every rational representation of Sp_g underlies a polarized $\mathbb{Z}$ variation of Hodge structure over $\mathcal{M}_g(L)$.

LEMMA 8.1. *If $\mathbb{V} \to \mathcal{M}^n_{g,r}(L)$ is a good variation of Hodge structure of negative weight whose monodromy representation*

$$\Gamma^n_{g,r}(L) \to \mathrm{Aut}\, V_\circ \otimes \mathbb{Q}$$

factors through a rational representation of Sp_g and contains no copies of the trivial representation, the group of normal functions $s : \mathcal{M}^n_{g,r}(L) \to J\mathcal{V}$ is finitely *generated of rank bounded by*

$$\dim H^1(\Gamma^n_{g,r}(L), V_{\mathbb{Z}}).$$

PROOF. A normal function corresponds to a variation of mixed Hodge structure whose underlying local system is an extension

$$0 \to \mathbb{V} \to \mathbb{E} \to \mathbb{Z} \to 0$$

of the trivial local system by $\mathbb{V}$.

One can form the semidirect product $\Gamma^n_{g,r}(L) \ltimes V_{\mathbb{Z}}$, where the mapping class group acts on $V_{\mathbb{Z}}$ via a representation $L \to \operatorname{Aut} V$. The monodromy representation of the local system $\mathbb{E}$ gives a splitting

$$\rho : \Gamma^n_{g,r}(L) \to \Gamma^n_{g,r}(L) \ltimes V_{\mathbb{Z}}$$

of the natural projection

$$\Gamma^n_{g,r}(L) \ltimes V_{\mathbb{Z}} \to \Gamma^n_{g,r}(L). \tag{4}$$

The splitting is well defined up to conjugation by an element of $V_{\mathbb{Z}}$.

The first step in the proof is to show that an extension of $\mathbb{Q}$ by $\mathbb{V}$ in the category of $\mathbb{Q}$ variations of mixed Hodge structure is determined by its monodromy representation. Two such variations can be regarded as elements of the group

$$\operatorname{Ext}^1_{\mathcal{H}(\mathcal{M}^n_{g,r}(L))}(\mathbb{Q}, \mathbb{V}). \tag{5}$$

It is easily seen that their difference is an extension whose monodromy representation factors through the homomorphism $\Gamma^n_{g,r} \to \mathrm{Sp}_g(\mathbb{Q})$. It now follows from Proposition 9.2 and the assumption that $\mathbb{V}$ contain no copies of the trivial representation that this difference is the trivial element of (5). The assertion follows.

From [35, p. 106] it follows that the set of splittings of (4), modulo conjugation by elements of $V_{\mathbb{Z}}$, is isomorphic to

$$H^1(\Gamma^n_{g,r}(L), V_{\mathbb{Z}}).$$

It follows from Proposition 5.2 that this group is finitely generated. Since normal functions are determined by their monodromy, the result follows. □

If $\mathbb{V}$ contains the trivial representation, the group of normal functions is an uncountably generated divisible group. For example, if $\mathbb{V}$ has trivial monodromy, then all such extensions are pulled back from a point. The set of normal functions is then

$$\operatorname{Ext}^1_{\mathcal{H}}(\mathbb{Z}, V_o) \approx JV_o,$$

where V_o denotes the fiber over the base point.

THEOREM 8.2. *If, in addition, the fiber over the base point is an irreducible* Sp_g*-module with highest weight* λ *and Hodge weight* m*, then the group of normal functions* $s : \mathcal{M}_{g,r}^n(L) \to J\mathcal{V}$ *is finitely generated of rank*

$$\dim H^1(\Gamma_{g,r}^n(L), V(\lambda)) = \begin{cases} 1 & \text{if } \lambda = \lambda_3 \text{ and } m = -1; \\ r+n & \text{if } \lambda = \lambda_1 \text{ and } m = -1; \\ 0 & \text{otherwise.} \end{cases}$$

The upper bounds for the rank of the group of normal functions follow from Lemma 8.1, Proposition 5.2, and the fact that the monodromy representation associated to a normal function has to be a morphism of variations of mixed Hodge structure Proposition 9.3. It remains to show that these upper bounds are achieved. We do this by explicitly constructing normal functions.

Multiples of the generators mod torsion of the normal functions associated to $V(\lambda_1)$ can be pulled back from $\mathcal{M}_g^1(L)$ along the $n+r$ forgetful maps $\mathcal{M}_{g,r}^n(L) \to \mathcal{M}_g^1(L)$. There the normal function can be taken to be the one that takes (C, x) to the point $(2g-2)x - \kappa_C$ of $\mathrm{Pic}^0 C$, where κ_C denotes the canonical class of C.

A multiple of the normal function associated to λ_3 can be pulled back from $\mathcal{M}_g(L)$ along the forgetful map $\mathcal{M}_{g,r}^n(L) \to \mathcal{M}_g(L)$. We will describe how this normal function over $\mathcal{M}_g(L)$ arises geometrically. If C is a smooth projective curve of genus g and $x \in C$, we have the Abel–Jacobi mapping

$$\nu_x : C \to \mathrm{Jac}\, C.$$

Denote the algebraic one-cycle $\nu_{x*}C$ in $\mathrm{Jac}\, C$ by C_x. Denote the cycle $i_* C_x$ by C_x^-, where $i : \mathrm{Jac}\, C \to \mathrm{Jac}\, C$ takes u to $-u$. The cycle $C_x - C^-$ is homologous to zero, and therefore defines a point $\tilde{e}(C, x)$ in $JH_3(\mathrm{Jac}\, C, \mathbb{Z}(-1))$. Pontrjagin product with the class of C induces a homomorphism

$$A : \mathrm{Jac}\, C \to JH_3(\mathrm{Jac}\, C, \mathbb{Z}(-1)).$$

Denote the cokernel of A by $JQ(\mathrm{Jac}\, C)$. It is not difficult to show that

$$\tilde{e}(C, x) - \tilde{e}(C, y) = A(x - y).$$

It follows that the image of $\tilde{e}(C, x)$ in $JQ(\mathrm{Jac}\, C)$ is independent of x. The image will be denoted by $e(C)$.

The primitive decomposition

$$H_3(\mathrm{Jac}\, C, \mathbb{Q}) = H_1(\mathrm{Jac}\, C, \mathbb{Q}) \oplus PH_3(\mathrm{Jac}\, C, \mathbb{Q})$$

is the decomposition of $H_3(\mathrm{Jac}\, C)$ into irreducible Sp_g-modules, the highest weights of the pieces being λ_1 and λ_3, respectively.

Fix a level L so that $\mathcal{M}_g(L)$ is smooth. The union of the $JQ(\operatorname{Jac} C)$ forms the bundle $\mathcal{J}_{\lambda_3}$ of intermediate jacobians over $\mathcal{M}_g(L)$ associated to the variation of Hodge structure of weight -1 and monodromy the third fundamental representation $V(\lambda_3)$ of Sp_g.

THEOREM 8.3. *The section e of $\mathcal{J}_{\lambda_3}$ is a normal function. Every other normal function associated to this bundle is, up to torsion, a half-integer multiple of e.*

PROOF. This result is essentially proved in [19]. We give a brief sketch.

To see that e is a normal function, consider the bundle of intermediate jacobians $JH_3(\operatorname{Jac} C, \mathbb{Z}(-1))$ over $\mathcal{M}_g^1(L)$. It follows from Remark 6.2 that $(C, x) \mapsto \tilde{e}(C, x)$ is a normal function. The argument of [19, p. 97] shows that there is a canonical quotient of the variation corresponding to $\tilde{e}$. (It is the extension E in [19, display 10].) This variation does not depend on the base point x, and is therefore constant along the fibers of $\mathcal{M}_g^1(L) \to \mathcal{M}_g(L)$. It follows that this quotient variation is the pullback of a variation on $\mathcal{M}_g(L)$. This quotient variation is classified by e. It follows that e is a normal function.

Each normal function f associated to this bundle of intermediate jacobians induces an L-equivariant homomorphism

$$f_* : H_1(T_g, \mathbb{Z}) \to H_1(JQ, \mathbb{Z}) \approx \Lambda^3 H_1(C, \mathbb{Z})/H_1(C, \mathbb{Z}).$$

It follows from a monodromy computation in [18, (4.3.5)] (see also [19, (6.3)]) that e_* is twice the Johnson homomorphism

$$\tau_g : H_1(T_g, \mathbb{Z}) \to \Lambda^3 H_1(C, \mathbb{Z})/H_1(C, \mathbb{Z}).$$

Since this homomorphism is primitive—that is, not a non-trivial integral multiple of another such normal function—all other normal functions associated to λ_3 must have monodromy representations that are half-integer multiples of that of e. As we have seen in the proof of Lemma 8.1, such normal functions are determined, up to torsion, by their monodromy representation. The result follows. □

I don't know how to realize $e/2$ as a normal function in this sense. But I do know to construct a more general kind of normal function associated to the one-cycle C in $\operatorname{Jac} C$ that does realize $e/2$. It is a section of a bundle whose fiber over C is a principal $JQ(\operatorname{Jac} C)$ bundle. The details may be found in [19, p. 92].

REMARK 8.4. Using the results in Section 9 and Theorem 8.2, one can easily show that the rank of the group of normal functions in the theorem above is

$$\dim \Gamma \operatorname{Hom}_{\mathrm{Sp}_g(\mathbb{Q})}(H_1(T_{g,r}^n, \mathbb{Q}), V_{\mathbb{Q},C}),$$

where $H_1(T_{g,r}^n)$ is given the Hodge structure of weight -1 described in §9.

9. Technical results on variations over $\mathcal{M}_g$

In this section we prove several technical facts about variations of mixed Hodge structure over moduli spaces of curves that were used in Section 8. Throughout we will assume that L has been chosen so that $\Gamma_{g,r}^n(L)$ is torsion-free.

PROPOSITION 9.1. *The local system $\mathbb{V}(\lambda)$ over $\mathcal{M}_{g,r}^n(L)$ associated to the irreducible representation of Sp_g with highest weight λ underlies a good $\mathbb{Q}$ variation of (mixed) Hodge structure, and this variation is unique up to Tate twist.*

PROOF. First observe that the local system $\mathbb{H}$ corresponding to the fundamental representation $V(\lambda_1)$ occurs as a variation of Hodge structure over $\mathcal{M}_{g,r}^n(L)$ of weight 1; it is simply the local system $R^1\pi_*\mathbb{Q}$ associated to the universal curve $\mathcal{C} \to \mathcal{M}_{g,r}^n(L)$. The existence of the structure of a good variation of Hodge structure on the local system corresponding the the Sp_g-module with highest weight λ now follows using Weyl's construction of the irreducible representations of Sp_g—see, for example, [13, §17.3].

To prove uniqueness, suppose that $\mathbb{V}$ and $\mathbb{V}'$ are both good variations of mixed Hodge structure corresponding to the same irreducible Sp_g-module. From Saito [43], we know that

$$\mathrm{Hom}_{\Gamma_{g,r}^n(L)}(\mathbb{V}, \mathbb{V}')$$

has a mixed Hodge structure. By Schur's lemma, this group is one-dimensional. It follows that this group is isomorphic to $\mathbb{Q}(n)$ for some n, and therefore that $\mathbb{V}' = \mathbb{V}(n)$. □

PROPOSITION 9.2. *If $\mathbb{E}$ is a good variation of $\mathbb{Q}$ mixed Hodge structure over $\mathcal{M}_g(L)$ whose monodromy representation factors through a rational representation of the algebraic group Sp_g, then for each dominant integral weight λ of Sp_g, the λ-isotypical part $\mathbb{E}_\lambda$ of $\mathbb{E}$ is a good variation of mixed Hodge structure. Consequently,*

$$\mathbb{E} = \bigoplus_\lambda \mathbb{E}_\lambda$$

in the category of good variations of $\mathbb{Q}$ mixed Hodge structure over $\mathcal{M}_g(L)$. Moreover, for each λ, there is a mixed Hodge structure A_λ such that $\mathbb{E}_\lambda = A_\lambda \otimes \mathbb{V}(\lambda)$.

PROOF. Fix λ, and let $\mathbb{V}(\lambda) \to \mathcal{M}_g(L)$ be a variation of Hodge structure whose fiber over some fixed base point is the irreducible Sp_g-module with highest weight λ. It follows from Saito's work [43] that

$$A_\lambda := \mathrm{Hom}_{\Gamma_g(L)}\big(\mathbb{V}(\lambda), \mathbb{E}\big) = H^0\big(\mathcal{M}_g(L), \mathrm{Hom}_{\mathbb{Q}}(\mathbb{V}(\lambda), \mathbb{E})\big)$$

is a mixed Hodge structure. Let

$$\mathbb{E}' = \bigoplus_\lambda A_\lambda \otimes \mathbb{V}(\lambda).$$

This is a good variation of mixed Hodge structure that is isomorphic to $\mathbb{E}$ as a $\mathbb{Q}$ local system. Now

$$\operatorname{Hom}_{\Gamma_g(L)}(\mathbb{E}',\mathbb{E}) = \bigoplus_\lambda A_\lambda^* \otimes \operatorname{Hom}_{\Gamma_g(L)}(\mathbb{V}(\lambda),\mathbb{E}) = \bigoplus_\lambda \operatorname{Hom}_{\mathbb{Q}}(A_\lambda, A_\lambda).$$

The element of this group that corresponds to $\mathrm{id} : A_\lambda \to A_\lambda$ in each factor is an isomorphism of local systems and an element of

$$\Gamma \operatorname{Hom}_{\Gamma_g(L)}(\mathbb{E}',\mathbb{E}).$$

It is therefore an isomorphism of variations of mixed Hodge structure. □

Now suppose that $g \geq 3$. The local system

$$\{H_1(T_{g,r}^n)\}$$

over $\mathcal{M}_{g,r}^n(L)$ naturally underlies a variation of mixed Hodge structure of weight -1. The λ_1 isotypical component is simply $r+n$ copies of the variation $\mathbb{V}(\lambda_1)$. We shall denote this variation by $\mathbb{H}_1(T_{g,r}^n)$.

Proposition 9.3. *Suppose that $\mathbb{V}$ is a variation of mixed Hodge structure over $\mathcal{M}_{g,r}^n(L)$ whose monodromy representation factors through a rational representation of Sp_g. If $\mathbb{E}$ is an extension of $\mathbb{Q}$ by $\mathbb{V}$ in the category of variations of mixed Hodge structure over $\mathcal{M}_{g,r}^n(L)$, then the restriction of the monodromy representation to $H_1(T_{g,r}^n)$,*

$$\mathbb{H}_1(T_{g,r}^n) \to \mathbb{V},$$

is a morphism of variations of mixed Hodge structure.

Proof. It suffices to prove the assertion for $\mathbb{Q}$ variations of mixed Hodge structure. We will prove the case $n = r = 0$, the proofs of the other cases being similar.

If the monodromy representation of $\mathbb{E}$ is trivial, the result is trivially true. So we shall assume that the monodromy representation is non-trivial.

Using the previous result, we can write

$$\mathbb{V} = \bigoplus_\lambda \mathbb{V}_\lambda$$

as variations of mixed Hodge structure over $\mathbb{V}$. By pushing out the extension

$$0 \to \mathbb{V} \to \mathbb{E} \to \mathbb{Q} \to 0$$

along the projection $\mathbb{V} \to \mathbb{V}_{\lambda_3}$ onto the λ_3 isotypical component, we obtain an extension

$$0 \to \mathbb{V}_{\lambda_3} \to \mathbb{E}' \to \mathbb{Q} \to 0.$$

It follows from Johnson's computation that the restricted monodromy representation of $\mathbb{E}$ factors through that of $\mathbb{E}'$:

$$\mathbb{H}_1(T_g) \to \mathbb{V}_{\lambda_3} \to \mathbb{V}.$$

We have therefore reduced to the case where $\mathbb{V} = \mathbb{V}_{\lambda_3}$.

Let $\mathbb{V}(\lambda_3)$ be the unique variation of Hodge structure of weight -1 over $\mathcal{M}_g(L)$ with monodromy representation given by λ_3. Let $\mathbb{S}$ be the variation of mixed Hodge structure over $\mathcal{M}_g(L)$ given by the cycle $C - C^-$ that was constructed in Section 8. It is an extension of $\mathbb{Q}$ by $\mathbb{V}(\lambda_3)$.

By [43], the exact sequence

$$0 \to \mathrm{Hom}_{\Gamma_g(L)}(\mathbb{S}, \mathbb{V}_{\lambda_3}) \to \mathrm{Hom}_{\Gamma_g(L)}(\mathbb{S}, \mathbb{E}') \to \mathrm{Hom}_{\Gamma_g(L)}(\mathbb{S}, \mathbb{Q})$$

is a sequence of mixed Hodge structures. The group on the right is easily seen to be isomorphic to $\mathbb{Q}(0)$; it is generated by the projection $\mathbb{S} \to \mathbb{Q}$. The group on the left is easily seen to be zero. It follows that

$$\mathrm{Hom}_{\Gamma_g(L)}(\mathbb{S}, \mathbb{E}') \approx \mathbb{Q}(0).$$

Since the monodromy representation of $\mathbb{S}$ is a morphism, so are those of $\mathbb{E}'$ and $\mathbb{E}$. □

10. Normal functions and cycles mod algebraic equivalence

As our first application of the classification of normal functions, we show that certain homologically trivial cycles defined over $\mathcal{M}^n_{g,r}(L)$ are of infinite order modulo algebraic equivalence for the general curve. We first recall a basic result, which follows from the fact that algebraic equivalences are parameterized by curves, and because the correspondence corresponding to a cycle algebraically equivalent to zero induces a map from the jacobian of the base curve to the appropriate intermediate jacobian of the ambient variety.

PROPOSITION 10.1. *Suppose that X is a smooth projective variety. If Z is a d-cycle that is algebraically equivalent to zero, the corresponding point $\nu(Z)$ of $J_d(X)$ lies in an abelian subvariety of $J_d(X)$.* □

LEMMA 10.2. *Suppose that $\mathbb{V} \to \mathcal{M}$ is a polarized variation of Hodge structure of weight -1. If the monodromy representation of $\mathbb{V}$ is irreducible, then either $J\mathcal{V}$ is a family of abelian varieties, or else the set*

$$\{m \in \mathcal{M} : JV_m \text{ contains an abelian variety of positive dimension}\}$$

is a countable union of proper subvarieties of $\mathcal{M}$.

PROOF. There are only a countable number of orthogonal decompositions

$$V_{o,\mathbb{Q}} = A \oplus B$$

of the fiber over the base point $o \in \mathcal{M}$. For each such decomposition there is the idempotent $p_A \in \operatorname{End} V_o$, which is orthogonal projection onto A. This is, in general, a multivalued section of the local system $\operatorname{End}_{\mathbb{Q}} \mathbb{V}$. The locus of the $m \in \mathcal{M}$ over which the splitting holds in the category of Hodge structures is the locus over which p_A is a Hodge class. This is an analytic subvariety of $\mathcal{M}$. In the case where A is an abelian variety and this locus is all of $\mathcal{M}$, the irreducibility of the monodromy implies that A and all its parallel translates span V_o. Since A has level 1 as a Hodge structure, this implies that V_o, and therefore $\mathbb{V}$ also, has level 1. That is, $J\mathcal{V}$ is a family of abelian varieties. □

Now suppose that $g \geq 3$. As in the previous section, we shall denote the unique $\mathbb{Q}$-variation of Hodge structure over $\mathcal{M}_{g,r}^n(L)$ of weight -1 associated to $V(\lambda_3)$ by $\mathbb{V}(\lambda_3)$. We shall denote the fiber over $C \in \mathcal{M}_{g,r}^n(L)$ of a family $W \to \mathcal{M}_{g,r}^n(L)$ by W_C.

THEOREM 10.3. *Suppose that $\pi_* : X \to \mathcal{M}_{g,r}^n(L)$ is a family of projective varieties, smooth over the generic curve, and that Z is a family of homologically trivial algebraic d-cycles in X defined generically over $\mathcal{M}_{g,r}^n(L)$. If the local system $R^{2d+1}\pi_*\mathbb{Q}_X(d+1)$ contains a copy of the variation $\mathbb{V}(\lambda_3)$, and if the component of the normal function of Z in the corresponding bundle of intermediate jacobians $\mathcal{J}_{\lambda_3}$ is of infinite order, then, for the general curve C, the cycle Z_C has infinite order modulo algebraic equivalence.*

PROOF. By Theorem 7.1, the normal function of Z is defined over all of the moduli space. Since the λ_3-component ν of this normal function has infinite order, and since $\mathcal{J}_{\lambda_3}$ is not a family of abelian varieties, it follows from Lemma 10.2 that, for the general curve, $\nu(C)$ is of infinite order modulo the maximal abelian subvariety of $JV_{\lambda_3,C}$. The result follows. □

Now take X to be the Jacobian of the universal curve over $\mathcal{M}_g^1(L)$, and Z to be the cycle whose fiber over $(C,x) \in \mathcal{M}_g^1$ is $C_x^{(a)} - i_*C_x^{(a)}$. Here

$$C_x^{(a)} := \{x_1 + \cdots + x_a - ax : x_j \in C\} \subseteq \operatorname{Jac} C$$

and i is the involution $D \mapsto -D$ of the jacobian. Applying the previous result and [19, (8.8)] we obtain the following result of Ceresa [10].

COROLLARY 10.4. *For the general curve C of genus g and $g \geq 3$, the cycle $C_x^{(a)} - i_*C_x^{(a)}$ is of infinite order modulo algebraic equivalence when $1 \leq a < g-1$.* □

When $g = 2$, the cycle $C_x - i_* C_x$ is algebraically equivalent to zero because, mod algebraic equivalence, we may assume x to be a Weierstrass point. In this case, the cycle is actually zero.

11. The Harris–Pulte theorem

As an application of the classification of normal functions above, we give a new proof of the Harris–Pulte theorem, which relates the mixed Hodge structure on $\pi_1(C,x)$ to the normal function of the cycle $C_x - C_x^-$ when $g \geq 3$. The result we obtain is slightly stronger.

Fix a level so that $\Gamma_g^1(L)$ is torsion-free. Denote by $\mathbb{L}$ the $\mathbb{Z}$ variation of Hodge structure of weight -1 over $\mathcal{M}_g^1(L)$ whose fiber over the pointed curve (C,x) is $H_1(C)$. Denote the corresponding holomorphic vector bundle by $\mathcal{L}$. The cycle $C_x - C_x^-$ defines a normal function ζ, which is a section of

$$J\Lambda^3\mathcal{L} \to \mathcal{M}_g^1(L).$$

Denote the integral group ring of $\pi_1(C,x)$ by $\mathbb{Z}\pi_1(C,x)$, and its augmentation ideal by $I(C,x)$, or I when there is no possibility of confusion. There is a canonical mixed Hodge structure on the truncated augmentation ideal

$$I(C,x)/I^3.$$

(See, for example, [16].) It is an extension

$$0 \to H_1(C)^{\otimes 2}/q \to I(C,x)/I^3 \to H_1(C) \to 0,$$

where q denotes the symplectic form. Tensoring with $H_1(C)$ and pulling back the resulting extension along the map $\mathbb{Z} \to H_1(C)^{\otimes 2}$, we obtain an extension

$$0 \to H_1(C) \otimes \left(H_1(C)^{\otimes 2}/q\right) \to E(C,x) \to \mathbb{Z} \to 0.$$

Since the set of $I(C,x)$ form a good variation of mixed Hodge structure over $\mathcal{M}_g^1(L)$ [17], the set of $E(C,x)$ form a good variation of mixed Hodge structure $\mathbb{E}$ over $\mathcal{M}_g^1(L)$. It therefore determines a normal function ρ, which is a section of

$$J\mathcal{L} \otimes \left(\mathcal{L}^{\otimes 2}/q\right) \to \mathcal{M}_g^1(L).$$

Define the map

$$\Phi : J\Lambda^3\mathcal{L} \to J\mathcal{L} \otimes (\mathcal{L}^{\otimes 2}/q)$$

to be the one induced by the map

$$\Lambda^3\mathbb{L} \to \mathbb{L}^{\otimes 3} \to \mathbb{L} \to \mathbb{L} \otimes (\mathbb{L}^{\otimes 2}/q);$$

the first map is defined by

$$x_1 \wedge x_2 \wedge x_3 \mapsto \sum_\sigma \operatorname{sgn}(\sigma) x_{\sigma(1)} \otimes x_{\sigma(2)} \otimes x_{\sigma(3)}$$

where σ ranges over all permutations of $\{1,2,3\}$.

Our version of the Harris–Pulte Theorem is:

THEOREM 11.1. *The image of ζ under Φ is 2ρ.*

PROOF. The proof uses Corollary 6.4. It is a straightforward consequence of Corollary 4.5 that the monodromy representations of $\Phi(\zeta)$ and 2ρ are equal. It is also a straightforward matter to use functoriality to show that both $\Phi(\zeta)$ and 2ρ vanish at (C,x) when C is hyperelliptic and x is a Weierstrass point [16, (7.5)].

□

12. The Franchetta conjecture for curves with a level

Suppose that L is a finite-index subgroup of $\mathrm{Sp}_g(\mathbb{Z})$, not necessarily torsion-free. Denote the generic point of $\mathcal{M}_g(L)$ by η. There is a universal curve defined generically over $\mathcal{M}_g(L)$. Denote its fiber over η by $\mathcal{C}_g(L)_\eta$. In the statement below, S denotes a compact oriented surface of genus g.

THEOREM 12.1. *For all $g \geq 3$ and all finite-index subgroups L of $\mathrm{Sp}_g(\mathbb{Z})$, the group $\mathrm{Pic}\,\mathcal{C}_g(L)_\eta$ is finitely generated of rank one. The torsion subgroup is isomorphic to $H^0(L, H_1(S,\mathbb{Q}/\mathbb{Z}))$. Modulo torsion, either it is generated by the canonical bundle, or by a divisor of degree $g-1$.*

This has a concrete statement when $L = \mathrm{Sp}_g(\mathbb{Z})(l)$, the congruence subgroup of level l of $\mathrm{Sp}_g(\mathbb{Z})$. It is not difficult to show that the only torsion points of $\mathrm{Jac}\, S$ invariant under L are the points of order l. That is,

$$H^0(L, H_1(S,\mathbb{Q}/\mathbb{Z})) \approx H_1(S,\mathbb{Z}/l\mathbb{Z}).$$

In this case we shall denote $\mathcal{C}_g(L)_\eta$ by $\mathcal{C}_g(l)_\eta$. During the proof of the theorem, we will show that, mod torsion, $\mathrm{Pic}\,\mathcal{C}_g(l)_\eta$ is generated by a theta characteristic when l is even. Combining this with the theorem, we have:

COROLLARY 12.2. *If $g \geq 3$, then for all $l \geq 0$, $\mathrm{Pic}\,\mathcal{C}_g(l)_\eta$ is a finitely generated group of rank one with torsion subgroup isomorphic to $H_1(S,\mathbb{Z}/l\mathbb{Z})$. Modulo torsion, $\mathrm{Pic}\,\mathcal{C}_g(l)_\eta$ is generated by a theta characteristic when l is even, and by the canonical bundle when l is odd.* □

The case $g = 2$, if true, should follow from Mess's computation of $H_1(T_2)$ [37]. One should note that Mess proved that T_2 is a countably generated free group.

SKETCH OF PROOF OF THEOREM 12.1. We first suppose that L is torsion-free. In this case, the universal curve is defined over all of $\mathcal{M}_g(L)$. Denote the restriction of it to a Zariski open subset U of $\mathcal{M}_g(L)$ by $\mathcal{C}_g(L)_U$. Set

$$\operatorname{Pic}_{\mathcal{C}_g/U} \mathcal{C}_g(L) = \operatorname{coker}\{\operatorname{Pic} U \to \operatorname{Pic} \mathcal{C}_g(U)\}.$$

Then

$$\operatorname{Pic} \mathcal{C}_g(L)_\eta = \varinjlim_{U} \operatorname{Pic}_{\mathcal{C}_g/U} \mathcal{C}_g(L),$$

where U ranges over all Zariski open subsets of $\mathcal{M}_g(L)$. There is a natural homomorphism

$$\deg : \operatorname{Pic}_{\mathcal{C}_g/U} \mathcal{C}_g \to \mathbb{Z}$$

given by taking the degree on a fiber. Denote $\deg^{-1}(d)$ by $\operatorname{Pic}^d_{\mathcal{C}_g/U} \mathcal{C}_g(L)$.

We first compute $\operatorname{Pic}^0 \mathcal{C}_g(L)_\eta$. Each element of this group can be represented by a line bundle over $\mathcal{C}_g(L)_U$ whose restriction to each fiber of $\pi : \mathcal{C}_g(L)_U \to U$ is topologically trivial. This line bundle has a section. By tensoring it with the pullback of a line bundle on U, if necessary, we may assume that the divisor of this section intersects each fiber of π in only a finite number of points. We therefore obtain a normal function

$$s : U \to \operatorname{Pic}^0_{\mathcal{C}_g/U} \mathcal{C}_g(L).$$

Since the associated variation of Hodge structure is the unique one of weight -1 associated to $V(\lambda_1)$, it follows from Theorem 8.2 and Theorem 7.1 that this normal function is torsion. It follows that

$$\operatorname{Pic}^0 \mathcal{C}_g(L)_\eta = \operatorname{Pic}^0_{\mathcal{C}_g/U} \mathcal{C}_g(L) = H^0(L, H_1(S, \mathbb{Q}/\mathbb{Z})).$$

Since this group is isomorphic to $H_1(S, \mathbb{Z}/l\mathbb{Z})$ when L is the congruence l subgroup of $\operatorname{Sp}_g(\mathbb{Z})$, and since every finite-index subgroup of $\operatorname{Sp}_g(\mathbb{Z})$ contains a congruence subgroup by [3], it follows that $\operatorname{Pic}^0 \mathcal{C}_g(L)_\eta$ is finite for all L.

The relative dualizing sheaf ω of $\mathcal{C}_g(L)_U$ gives an element of $\operatorname{Pic}^{2g-2} \mathcal{C}_g(L)_\eta$. Denote the greatest common divisor of the degrees of elements of $\operatorname{Pic} \mathcal{C}_g(L)_\eta$ by d. Observe that d divides $2g-2$. Let $m = (2g-2)/d$. We will show that $m = 1$ or 2.

Choose an element δ of $\operatorname{Pic}^d \mathcal{C}_g(L)_\eta$. Then

$$\omega - m\delta \in \operatorname{Pic}^0 \mathcal{C}_g(L)_\eta$$

and is therefore a torsion element of order k, say. Replace L by

$$L' = L \cap \operatorname{Sp}_g(\mathbb{Z})(km).$$

Observe that the natural map

$$\operatorname{Pic}^0 \mathcal{C}_g(L)_\eta \to \operatorname{Pic}^0 \mathcal{C}_g(L')_\eta$$

is injective. We can find

$$\mu \in \operatorname{Pic}^0 \mathcal{C}_g(L')_\eta$$

such that $m\mu = \omega - m\delta$. Then $\delta + \mu$ is an m-th root of the canonical bundle ω. It follows from a result of Sipe [46] that the only non-trivial roots of the canonical bundle that can be defined over $\mathcal{M}_g(L)$ are square roots: see Theorem 13.3. This implies that m divides 2, as claimed.

It also follows from Theorem 13.3 that square roots of the canonical bundle are defined over $\mathcal{M}_g(l)$ if and only if l is even. Combined with the argument above, this shows that, mod torsion, $\operatorname{Pic}^0 \mathcal{C}_g(l)_\eta$ is generated by ω if l is odd, and by a square root of ω if l is even.

Our final task is to reduce the general case to that where L is torsion-free. For arbitrary L, we have

$$\operatorname{Pic} \mathcal{C}_g(L)_\eta = \varinjlim_{U} \operatorname{Pic}_{\mathcal{C}_g/U} \mathcal{C}_g(L),$$

where U ranges over all smooth Zariski open subsets of $\mathcal{M}_g(L)$. Choose a torsion-free finite-index normal subgroup L' of L and a smooth Zariski open subset U of $\mathcal{M}_g(L)$. Denote the inverse image of U in $\mathcal{M}_g(L')$ by U'. Then the projection $U' \to U$ is a Galois cover with Galois group $G = L/L'$. It follows that

$$\operatorname{Pic}_{\mathcal{C}_g/U} \mathcal{C}_g(L) = \operatorname{Pic}_{\mathcal{C}_g/U'} \mathcal{C}_g(L')^G.$$

Since $\pi_1(U)$ surjects onto $\Gamma_g(L)$, and therefore onto L, the result follows. □

Denote the universal curve over the generic point η of $\mathcal{M}^n_{g,r}(l)$ by $\mathcal{C}^n_{g,r}(l)_\eta$. The proof of the following more general result is similar to that of Theorem 12.1.

THEOREM 12.3. *If $g \geq 3$, then for all $l \geq 0$, $\operatorname{Pic} \mathcal{C}^n_{g,r}(l)_\eta$ is a finitely generated group of rank $r+n+1$ whose torsion subgroup is isomorphic to $H_1(S, \mathbb{Z}/l\mathbb{Z})$. Each of the n marked points and the anchor point of each of the r marked cotangent vectors gives an element of $\operatorname{Pic}^1 \mathcal{C}^n_{g,r}(l)_\eta$. The pairwise differences of these points generate a subgroup of $\operatorname{Pic}^0 \mathcal{C}^n_{g,r}(l)_\eta$ of rank $r+n-1$. Moreover, $\operatorname{Pic}^0 \mathcal{C}^n_{g,r}(l)_\eta$ is generated by these differences modulo torsion, and $\operatorname{Pic} \mathcal{C}^n_{g,r}(l)_\eta$ is generated modulo $\operatorname{Pic}^0 \mathcal{C}^n_{g,r}(l)_\eta$ by the class of one of the distinguished points together with a theta characteristic when l is even, and by the canonical divisor when l is odd.* □

Note that the independence of the pairwise difference of the points follows from the discussion following Theorem 8.2.

13. The monodromy of roots of the canonical bundle

In this section we compute the action of Γ_g on the set of n-th roots of the canonical bundle of a curve of genus $g \geq 3$. This computation is a slight refinement of a result of P. Sipe [46].

If L is an n-th root of the tangent bundle of a smooth projective curve C, its dual is an n-th root of the canonical bundle. That is, there is a one-one correspondence between n-th roots of the canonical bundle and n-th roots of the tangent bundle of a curve. For convenience, we shall work with roots of the tangent bundle.

The first point is that roots of the tangent bundle are determined topologically (see [2, §3] and [46]): denote the $\mathbb{C}^*$ bundle associated to the holomorphic tangent bundle TC of C by T^*. Indeed, an n-th root of TC is a cyclic covering of T^* of degree n, which has degree n on each fiber. The complex structure on such a covering is uniquely determined by that on T^*.

The first Chern class of TC is $2-2g$. So if R is an n-th root of K, the integer n divides $2g-2$. Since the Euler class of T^* is $2-2g$, it follows from the Gysin sequence that there is a short exact sequence

$$0 \to \mathbb{Z}/n\mathbb{Z} \to H_1(T^*, \mathbb{Z}/n\mathbb{Z}) \to H_1(C, \mathbb{Z}/n\mathbb{Z}) \to 0. \tag{6}$$

By covering space theory, an n-th root of TC is determined by a homomorphism

$$H_1(T^*, \mathbb{Z}/n\mathbb{Z}) \to \mathbb{Z}/n\mathbb{Z}$$

whose composition with the inclusion $\mathbb{Z}/n\mathbb{Z} \hookrightarrow H_1(T^*, \mathbb{Z}/n\mathbb{Z})$ is the identity. That is, we have the following result:

PROPOSITION 13.1. *There is a natural one-to-one correspondence between n-th roots of the the canonical bundle of C and splittings of the sequence* (6).

□

Throughout this section, we will assume $g \geq 3$. Denote the set of n-th roots of TC by Θ_n. This is a principal $H_1(C, \mathbb{Z}/n\mathbb{Z})$ space. The automorphism group of this affine space is an extension

$$0 \to H_1(C, \mathbb{Z}/n\mathbb{Z}) \to \operatorname{Aut}\Theta_n \xrightarrow{\pi} GL_{2g}(\mathbb{Z}/n\mathbb{Z}) \to 1;$$

the kernel being the group of translations by elements of $H_1(C, \mathbb{Z}/n\mathbb{Z})$. The mapping class group acts on Θ_n, so we have a homomorphism

$$\Gamma_g \to \operatorname{Aut}\Theta_n.$$

The composite of this homomorphism with π is the reduction mod n

$$\rho_n : \Gamma_g \to \mathrm{Sp}_g(\mathbb{Z}/n\mathbb{Z})$$

of the natural homomorphism. Denote the subgroup $\pi^{-1}(\mathrm{Sp}_g(\mathbb{Z}/n\mathbb{Z}))$ of $\operatorname{Aut}\Theta_n$ by $\mathcal{K}_n$. It follows that the action of Γ_g on Θ_n factors through a homomorphism

$$\theta_n : \Gamma_g \to \mathcal{K}_n$$

whose composition with the natural projection $\mathcal{K}_n \to \mathrm{Sp}_g(\mathbb{Z}/n\mathbb{Z})$ is ρ_n. In order to determine θ_n, we will need to compute its restriction

$$\theta_n : H_1(T_g) \to H_1(C, \mathbb{Z}/n\mathbb{Z})$$

to the Torelli group. First some algebra.

PROPOSITION 13.2. *There is a natural homomorphism*

$$\psi_g : H_1(T_g, \mathbb{Z}) \to H_1(C,\ \mathbb{Z}/(g-1)\mathbb{Z}).$$

PROOF. By Theorem 3.4, there is a natural homomorphism

$$\tau_g : H_1(T_g, \mathbb{Z}) \to \Lambda^3 H_1(C, \mathbb{Z}) / \big([C] \times H_1(C, \mathbb{Z})\big).$$

Here we view $\Lambda^\bullet H_1(C)$ as the homology of $\operatorname{Jac} C$ and $[C]$ denotes the homology class of the image of C under the Abel–Jacobi map. There is also a natural homomorphism

$$p : \Lambda^3 H_1(C, \mathbb{Z}) \to H_1(C, \mathbb{Z})$$

defined by

$$p : x \wedge y \wedge z \mapsto (x \cdot y)\, z + (y \cdot z)\, x + (z \cdot x)\, y.$$

It is easy to see that the composite

$$H_1(C, \mathbb{Z}) \xrightarrow{[C]\times} \Lambda^3 H_1(C, \mathbb{Z}) \xrightarrow{p} H_1(C, \mathbb{Z})$$

is multiplication by $g-1$. It follows that p induces a homomorphism

$$\overline{p} : \Lambda^3 H_1(C, \mathbb{Z}) \to H_1(C,\ \mathbb{Z}/(g-1)\mathbb{Z}).$$

The homomorphism ψ_g is the composite $\overline{p} \circ \tau_g$. □

Call a translation of Θ_n *even* if it is translation by an element of

$$2H^1(C, \mathbb{Z}/n\mathbb{Z}).$$

If n is odd, this is the set of all translations. If $n = 2m$, this is the proper subgroup of $H^1(C, \mathbb{Z}/n\mathbb{Z})$ isomorphic to $H^1(C, \mathbb{Z}/m\mathbb{Z})$.

THEOREM 13.3. *The image of the natural homomorphism* $\theta_n : \Gamma_g \to \mathcal{K}_n$ *is a subgroup* $\mathcal{K}_n^{(2)}$ *of* $\mathcal{K}_n$, *which is an extension*

$$0 \to 2\, H_1(C, \mathbb{Z}/n\mathbb{Z}) \to \operatorname{Aut} \Theta_n \xrightarrow{\pi} \mathrm{GL}_{2g}(\mathbb{Z}/n\mathbb{Z}) \to 1$$

of $\mathrm{Sp}_g(\mathbb{Z}/n\mathbb{Z})$ *by the even translations. The restriction of* θ_n *to* T_g *is the composite of* ψ_g *with the homomorphism*

$$H_1(C,\, \mathbb{Z}/(g-1)\mathbb{Z}) \xrightarrow{r} H_1(C, \mathbb{Z}/n\mathbb{Z}) \xrightarrow{\mathrm{PD}} H^1(C, \mathbb{Z}/n\mathbb{Z}),$$

where $r(k)$ *equals* $2k$ *mod* n, *and* PD *denotes Poincaré duality. In particular, the Torelli group acts trivially on* Θ_n *if and only if* n *divides* 2.

PROOF. First, Johnson proved in [31] that the kernel of the composite

$$T_g \to H_1(T_g) \xrightarrow{\tau_g} \Lambda^3 H_1(C, \mathbb{Z})/([C] \times H_1(C, \mathbb{Z}))$$

is generated by Dehn twists on separating simple closed curves. Using this, it is easy to check that the restriction of θ_n to T_g factors through τ_g. In [32], Johnson shows that T_g is generated by Dehn twists on a bounding pair of disjoint simple closed curves. (Actually, all we need is that $\Lambda^3 H_1(C, \mathbb{Z})/([C] \times H_1(C, \mathbb{Z}))$ be generated by the images under τ_g by such bounding pair maps. This is easily checked directly.)

Now suppose that φ is such a bounding pair map. There are two disjoint embedded circles A and B such that φ equals a positive Dehn twist about A and a negative one about B. When we cut C along $A \cup B$, we obtain two surfaces, of genera g' and g'', say. Choose one of these components, and let a be the cycle obtained by orienting A so that it is a boundary component of this component. It is not difficult to show that the image of φ under ψ_g equals

$$-g'\,[a] \in H_1(C,\, \mathbb{Z}/(g-1)\mathbb{Z}),$$

where g' is the genus of the chosen component. Since $g' + g'' = g - 1$, this is well defined. Next, one can use Morse theory to show that the image of this same bounding pair map in $H^1(C, \mathbb{Z}/n\mathbb{Z})$ is $-2g'\,\mathrm{PD}(a)$. The result follows. □

COROLLARY 13.4. *The only roots of the canonical bundle defined over Torelli space are the canonical bundle itself and its* 2^{2g} *square roots.* □

REMARK 13.5. The homomorphism $\theta_{2g-2} : \Gamma_g \to \mathcal{K}_{2g-2}$ appears in Morita's work [38, §4.A].

14. Heights of Cycles defined over $\mathcal{M}_g(L)$

Suppose that X is a compact Kähler manifold of dimension n and that Z and W are two homologically trivial algebraic cycles in X of dimensions d and e, respectively. Suppose that $d+e=n-1$ and that Z and W have disjoint supports. Denote the current associated to W by δ_W. It follows from the $\partial\overline{\partial}$-lemma that there is a current η_W of type (d,d) that is smooth away from the support of Z and satisfies

$$\partial\overline{\partial}\eta_W = \pi i \delta_W.$$

The (archimedean) height pairing between Z and W is defined by

$$\langle Z, W\rangle = -\int_Z \eta_W.$$

This is a real-valued, symmetric bilinear pairing on such disjoint homologically trivial cycles. It is important in number theory [5].

Now suppose that

$$X \to \mathcal{M}_g(L)$$

is a family of smooth projective varieties of relative dimension n. Suppose that $Z \to \mathcal{M}_g(L)$ and $W \to \mathcal{M}_g(L)$ are families of algebraic cycles in X of relative dimensions d and e, respectively, where $d+e=n-1$. Denote the fiber of X, Z and W over $C \in \mathcal{M}_g(L)$ by X_C, Z_C and W_C, respectively. Suppose that Z_C and W_C are homologically trivial in X_C and that they have disjoint supports for generic $C \in \mathcal{M}_g(L)$.

We shall suppose that L has been chosen so that every curve has two distinguished theta characteristics α and $\alpha+\delta$, where δ is a non-zero point of order two in $\operatorname{Jac} C$. We shall also suppose that g is odd and ≥ 3. Write g in the form $g = 2d+1$.

Denote the difference divisor

$$\{x_1 + \cdots + x_d - y_1 - \cdots - y_d : x_j, y_j \in C\}$$

in $\operatorname{Jac} C$ by Δ, and the theta divisor

$$\{x_1 + \cdots + x_{2d} - \alpha : x_j \in C\}$$

in $\operatorname{Jac} C$ by Θ_α. By [18, (4,1.2)], there is a rational function f_C on $\operatorname{Jac} C$ whose divisor is

$$\Delta - \binom{2d}{d}\Theta_\alpha.$$

Denote the unique invariant measure of total mass one on $\operatorname{Jac} C$ by μ.

THEOREM 14.1. *Suppose that g is odd and ≥ 3. Suppose that Z and W are families of homologically trivial cycles over $\mathcal{M}_g(L)$ in a family of smooth projective varieties $p : X \to \mathcal{M}_g(L)$, as above. If the monodromy of the local system $R^{2d+1}p_*\mathbb{Q}_X$ factors through a rational representation of Sp_g, there is a rational function h on $\mathcal{M}_g(L)$ and rational numbers a and b such that*

$$\langle Z_C, W_C\rangle = a\Big(\log|h(C)| + 2b\Big(\log|f_C(\delta)| - \int_{\mathrm{Jac}\, C}\log|f_C(x)|d\mu(x)\Big)\Big).$$

The numbers a and b are topologically determined, as will become apparent in the proof. The divisor of h is computable when one has a good understanding of how the cycles Z and W intersect. One should be able to derive a similar formula for even g using Bost's general computation of the height in [8] and results from [19].

The proof of Theorem 14.1 occupies the remainder of this section. We only give a sketch. We commence by defining two algebraic cycles in $\mathrm{Pic}^d C$. For $D \in \mathrm{Jac}\, C$, let $C_D^{(d)}$ be the d-cycle in $\mathrm{Pic}^d C$ obtained by pushing forward the fundamental class of the d-th symmetric power of C along the map

$$\{x_1, \dots, x_d\} \mapsto x_1 + \cdots + x_d + D.$$

Let i be the automorphism of $\mathrm{Pic}^d C$ defined by $i : x \mapsto \alpha - x$. Define

$$Z_D = C_D^{(d)} - i_* C_D^{(d)}.$$

This is a homologically trivial d-cycle in $\mathrm{Pic}^d C$.

From [8] and [18], we know that

$$\langle Z_0, Z_\delta\rangle = 2\log|f_C(\delta)| - 2\int_{\mathrm{Jac}\, C}\log|f_C(x)|\, d\mu(x).$$

So the content of the theorem is that there is a rational function h on $\mathcal{M}_g(L)$ and rational numbers a and b such that

$$\langle Z, W\rangle = a\left(\log|h(C)| + b\langle Z_0, Z_\delta\rangle\right).$$

The basic point, as we shall see, is that, up to torsion, all normal functions over $\mathcal{M}_g(L)$ are half-integer multiples of that of $C - C^-$, as was proved in Section 8.

We will henceforth assume that the reader is familiar with the content of [18, §3]. We will briefly review the most relevant points of that section.

A *biextension* is a mixed Hodge structure B with only three non-trivial weight graded quotients: $\mathbb{Z}$, H, and $\mathbb{Z}(1)$, where H is a Hodge structure of weight -1. The isomorphisms

$$\mathrm{Gr}^W_{-2} B \approx \mathbb{Z}(1) \quad\text{and}\quad \mathrm{Gr}^W_0 B \approx \mathbb{Z}$$

are considered to be part of the data of the biextension. If one replaces $\mathbb{Z}$ by $\mathbb{R}$ in this definition, one obtains the definition of a real biextension. To a biextension

B one can associate a real number $\nu(B)$, called the *height* of B. It depends only on the associated real biextension $B \otimes \mathbb{R}$.

Associated to a pair of disjoint homologically trivial cycles in a smooth projective variety X satisfying

$$\dim Z + \dim W + 1 = \dim X$$

there is a canonical biextension $B_{\mathbb{Z}}(Z, W)$, whose weight graded quotients are

$$\mathbb{Z}, \quad H_{2d+1}(X, \mathbb{Z}(-d)), \quad \mathbb{Z}(1),$$

where d is the dimension of Z. The extensions

$$0 \to H_{2d+1}(X, \mathbb{Z}(-d)) \to B_{\mathbb{Z}}(Z, W)/\mathbb{Z}(1) \to \mathbb{Z} \to 0$$

and

$$0 \to \mathbb{Z}(1) \to W_{-1}B_{\mathbb{Z}}(Z, W) \to H_{2d+1}(X, \mathbb{Z}(-d)) \to 0$$

are the those determined by Z (directly), and W (via duality) [18, (3.3.2)]. We have

$$\nu(B_{\mathbb{Z}}(Z, W)) = \langle Z, W \rangle.$$

The first step in the proof is to reduce the size of the biextension. Suppose that $\Lambda = \mathbb{Z}$ or $\mathbb{R}$, and that B is a Λ-biextension with weight -1 graded quotient H. Suppose that there is an inclusion $i : A \hookrightarrow H$ of Λ mixed Hodge structures. Pulling back the extension

$$0 \to \Lambda(1) \to W_1 B \to H \to 0$$

along i, we obtain an extension

$$0 \to \Lambda(1) \to E \to C \to 0.$$

If this extension splits, there is a canonical lift $\tilde{\imath} : C \to B$ of i. The quotient B/C is also a Λ biextension.

PROPOSITION 14.2. *The biextensions* $B_\Lambda(Z, W)$ *and* $B_\Lambda(Z, W)/C$ *have the same height.*

PROOF. This is a special case of [34, (5.3.8)]. It follows directly from [18, (3.2.11)]. □

We will combine this with Theorem 8.2 to prune the biextension $B_{\mathbb{Z}}(Z_C, W_C)$ until its weight -1 graded quotient is either trivial or else one copy of $V(\lambda_3)$.

First observe that if B is a biextension and B' a mixed Hodge substructure of B of finite index, then B' is a biextension and there is a non-zero integer m such that $\nu(B') = m\nu(B)$. This can be proved using [18, (3.2.11)].

To prune the biextension $\mathbb{B}(Z,W)$ over $\mathcal{M}_g(L)$, we consider the portion of the monodromy representation

$$H_1(T_g) \to \mathrm{Hom}_{\mathbb{Z}}(\mathrm{Gr}^W_{-1} B(Z_C, W_C), \mathbb{Z}(1))$$

associated to the variation $W_{-1}\mathbb{B}(Z,W)$ over $\mathcal{M}_g(L)$. This map is Sp_g equivariant. Denote $\mathrm{Gr}^W_{-1}\mathbb{B}(Z,W)$ by $\mathbb{H}$. This monodromy representation corresponds to the map

$$\mathbb{H} \to \{H^1(T_g, \mathbb{Z}(1))\}$$

of local systems over $\mathcal{M}_g(L)$, which takes $h \in H_C$ to the functional $\{\phi \mapsto \phi(h)\}$ on $H_1(T_g)$. For each $C \in \mathcal{M}_g(L)$ this is a morphism of Hodge structures by Proposition 9.3. Denote its kernel by K_C. These form a variation of Hodge structure $\mathbb{K}$ over $\mathcal{M}_g(L)$. If the monodromy representation is trivial on T_g, then $\mathbb{K} = \mathbb{H}$. Otherwise, Schur's lemma implies that $\mathbb{H}/\mathbb{K}$ is isomorphic $\mathbb{V}(\lambda_3)$ placed in weight -1.

We can pull back the extension

$$0 \to \mathbb{Q}(1) \to W_{-1}\mathbb{B}(Z,W) \to \mathbb{H} \to 0$$

along the inclusion $\mathbb{K} \hookrightarrow \mathbb{H}$ to obtain an extension

$$0 \to \mathbb{Q}(1) \to \mathbb{E} \to \mathbb{K} \to 0. \tag{7}$$

If this extension splits over $\mathbb{Q}$, then, by replacing the lattice in $\mathbb{B}_{\mathbb{Z}}(Z,W)$ by a commensurable one, we may assume that the splitting is defined over $\mathbb{Z}$. This has the effect of multiplying the height by a non-zero rational number. Once we have done this, the inclusion $\mathbb{K} \hookrightarrow \mathrm{Gr}^W_{-1}\mathbb{B}(Z,W)$ lifts to an inclusion $\mathbb{K} \hookrightarrow \mathbb{B}(Z,W)$. Using Proposition 14.2, we can replace $B_{\mathbb{Z}}(Z_C, W_C)$ by $\mathbb{B}' = \mathbb{B}(Z,W)/\mathbb{K}$ without changing the height of the biextension. For the time being, we shall assume that (7) splits over $\mathbb{Q}$. This is the case, for example, when $\mathbb{H}$ contains no copies of the trivial representation, as follows from Proposition 9.2 since $\mathbb{K}$ is a trivial T_g-module by construction.

The weight -1 graded quotient of $\mathbb{B}'$ is either trivial or isomorphic to $\mathbb{V}(\lambda_3)$. This biextension is defined over the open subset U of $\mathcal{M}_g(L)$ where Z_C and W_C are disjoint. The related variations $W_{-1}\mathbb{B}'$ and $\mathbb{B}'/\mathbb{Z}(1)$ are defined over all of $\mathcal{M}_g(L)$.

If $\mathbb{K} = \mathbb{H}$, then $\mathbb{B}'$ is an extension of $\mathbb{Z}$ by $\mathbb{Z}(1)$. It therefore corresponds to a rational function h on $\mathcal{M}_g(L)$, which is defined on U [20, (9.3)]. It follows from [18, (3.2.11)] that the height of this biextension $\mathbb{B}'$ is $C \mapsto \log|h(C)|$. This completes the proof of the theorem in this case.

Dually, when the extension

$$0 \to \mathrm{Gr}^W_{-1}\mathbb{B}' \to \mathbb{B}'/\mathbb{Z}(1) \to \mathbb{Z} \to 0$$

has finite monodromy, there is a rational function h on $\mathcal{M}_g(L)$ such that the height of B', and therefore $B(Z_C, W_C)$, is rational multiple of $\log|h(C)|$.

We have therefore reduced to the case where $\mathbb{B}'$ has weight graded -1 quotient $\mathbb{V}(\lambda_3)$ and where neither of the extensions

$$0 \to \mathbb{V}(\lambda_3) \to \mathbb{B}'/\mathbb{Z}(1) \to \mathbb{Z} \to 0$$

or

$$0 \to \mathbb{Z}(1) \to W_{-1}\mathbb{B}' \to \mathbb{V}(\lambda_3) \to 0$$

is torsion. We also have the biextension $\mathbb{B}''$ associated to the cycles Z_0 and Z_δ. It has these same properties. After replacing the lattices in each by lattices of finite index, we may assume that the extensions of variations $W_{-1}\mathbb{B}'$ and $W_{-1}\mathbb{B}''$ are isomorphic, and that $\mathbb{B}'/\mathbb{Z}(1)$ and $\mathbb{B}''/\mathbb{Z}(1)$ are isomorphic. As in [18, (3.4)], the biextensions $\mathbb{B}'$ and $\mathbb{B}''$ each determine a canonically metrized holomorphic line bundle over $\mathcal{M}_g(L)$. These metrized line bundles depend only on the variations $\mathbb{B}/\mathbb{Z}(1)$ and $W_{-1}\mathbb{B}$, and are therefore isomorphic. Denote this common line bundle by $\mathcal{B} \to \mathcal{M}_g(L)$. The biextensions $\mathbb{B}'$ and $\mathbb{B}''$ determine (and are determined by) meromorphic sections s' and s'' of $\mathcal{B}$, respectively. There is therefore a meromorphic function h on $\mathcal{M}_g(L)$ such that $s'' = hs'$. It follows from the main result of [34] that this function is a rational function. (The philosophy is that period maps of variations of mixed Hodge structure behave well at the boundary.) The result follows as

$$\nu(B''_C) = \log\|s''(C)\| = \log|h(C)| + \log\|s'(C)\| = \nu(B'_C) + \log|h(C)|.$$

To conclude the proof, we now explain how to proceed when the extension (7) of page 137 is not split as a $\mathbb{Q}$ variation. Write $\mathbb{K} = \mathbb{T} \oplus \mathbb{T}'$, where $\mathbb{T}$ is the trivial submodule of $\mathbb{K}$ and $\mathbb{T}'$ is its orthogonal complement. This is a splitting in the category of $\mathbb{Q}$ variations by Proposition 9.2. It also follows from Proposition 9.2 that the restriction of (7) to $\mathbb{T}'$ is split. Consequently, there is an inclusion of mixed Hodge structures $\mathbb{T}' \hookrightarrow \mathbb{B}(Z, W)$. As above, we may replace $\mathbb{B}(Z, W)$ by the biextension $\mathbb{B}' = \mathbb{B}(Z, W)/\mathbb{T}'$ after rescaling lattices. This only changes the height by a non-zero rational number. The weight graded -1 quotient of $\mathbb{B}'$ is the sum of at most one copy of $\mathbb{V}(\lambda_3)$ and a trivial variation of weight -1.

Now suppose that B_1 and B_2 are two biextensions. We can construct a new biextension $B_1 \boxplus B_2$ from them as follows: Begin by taking their direct sum. Pull this back along the diagonal inclusion

$$\mathbb{Z} \hookrightarrow \mathbb{Z} \oplus \mathbb{Z} = \operatorname{Gr}_0^W (B_1 \oplus B_2)$$

to obtain a mixed Hodge structure B whose weight -2 graded quotient is

$$\operatorname{Gr}_{-2}^W (B_1 \oplus B_2) = \mathbb{Z}(1) \oplus \mathbb{Z}(1).$$

Push this out along the addition map

$$\mathbb{Z}(1) \oplus \mathbb{Z}(1) \to \mathbb{Z}(1)$$

to obtain the sought after biextension $B_1 \boxplus B_2$. The following result follows directly from [18, (3.2.11)].

PROPOSITION 14.3. *The height of $B_1 \boxplus B_2$ is the sum of the heights of B_1 and B_2.*

The biextension $\mathbb{B}'$ is easily seen to be the sum, in this sense, of two biextensions. The first is constant with weight -1 quotient equal to the trivial variation $\mathbb{T}$ and the second is a variation with weight -1 quotient equal to $\mathbb{H}/\mathbb{K}$, which is either zero or one copy of $\mathbb{V}(\lambda_3)$. Since the height of a constant biextension is a constant, the result follows from the computation of the height of a biextension with weight -1 quotient $\mathbb{V}(\lambda_3)$ above.

15. Results for Abelian Varieties

Denote by $\mathcal{A}_g(L)$ the quotient of Siegel space $\mathfrak{h}_g$ of rank g by a finite-index subgroup L of $\mathrm{Sp}_g(\mathbb{Z})$. This is the moduli space of abelian varieties with a level-L structure. In this section we state results for $\mathcal{A}_g(L)$ analogous to those in Sections 8 and 14. The proofs are similar, but much simpler, and are left to the reader.

We call a representation of Sp_g *even* if it has a symmetric Sp_g-invariant inner product, and *odd* if it has a skew symmetric Sp_g-invariant inner product. It follows from Schur's Lemma that every irreducible representation of Sp_g is either even or odd. The even ones occur as polarized variations of Hodge structure of even weight over each $\mathcal{A}_g(L)$, while the odd ones occur as polarized variations of Hodge structure only over $\mathcal{A}_g(L)$ of odd weight provided $-I \notin L$. These facts are easily proved by adapting the arguments in Section 9.

The first theorem is the analogue of Lemma 8.1 for abelian varieties. It is similar to the result [45] of Silverberg. The point in our approach is that $H^1(L, V)$ vanishes for all non-trivial irreducible representations of Sp_g by [42].

THEOREM 15.1. *Suppose that $g \geq 2$ and that $L/\pm I$ is torsion-free. If $\mathbb{V} \to \mathcal{A}_g(L)$ is a variation of Hodge structure of negative weight whose monodromy representation is the restriction to L of a rational representation of Sp_g, the group of generically defined normal functions associated to this variation is finite.* □

Since there are no normal functions of infinite order over $\mathcal{A}_g(L)$, we have the following analogue of Theorem 14.1. Suppose that Z and W are families of homologically trivial cycles over $\mathcal{A}_g(L)$ in a family of smooth projective varieties

$p : X \to \mathcal{A}_g(L)$. Suppose that they are disjoint over the generic point. Suppose further that $d + e = n - 1$, where d, e and n are the relative dimensions over $\mathcal{A}_g(L)$ of Z, W and X, respectively. Denote the fiber of Z over $A \in \mathcal{A}_g(L)$ by Z_A, etc.

THEOREM 15.2. *If $g \geq 2$ and the monodromy of the local system $R^{2d+1}p_*\mathbb{Q}_X$ is the restriction to L of a rational representation of* Sp_g, *there is a rational function h on $\mathcal{A}_g(L)$ such that*

$$\langle Z_A, W_A \rangle = \log |h(A)|$$

for all $A \in \mathcal{A}_g(L)$. □

One can formulate and prove analogues of these results for the moduli spaces $\mathcal{A}_g^n(L)$ of abelian varieties of dimension g, n marked points, and a level-L structure.

We conclude with a discussion of Nori's results and their relation to Theorems 8.2 and 15.1. We first recall the main result of the last section of [40].

THEOREM 15.3 (NORI). *Suppose that X is a variety that is an unbranched covering of a Zariski open subset U of $\mathcal{A}_g(L)$, where L is torsion-free. Suppose that $\mathbb{V}$ is a variation of Hodge structure of negative weight over X that is pulled back from the canonical variation over $\mathcal{A}_g(L)$ of the same weight whose monodromy representation is irreducible and has highest weight λ. Then the group of normal functions defined on X associated to this variation is finite unless*

$$\lambda = \begin{cases} 0 & \text{and } g \geq 2, \text{ or} \\ \lambda_1 & \text{and } g \geq 3, \text{ or} \\ \lambda_3 & \text{and } g = 3, \text{ or} \\ m_1\lambda_1 + m_2\lambda_2 & g = 2 \text{ and } m_1 \geq 2. \end{cases}$$

□

This result may seem to contradict Theorem 15.1. The difference can be accounted for by noting that Theorem 15.1 only applies to open subsets of the $\mathcal{A}_g(L)$, whereas Nori's theorem applies to a much more general class of varieties that contains unramified coverings of open subsets of the $\mathcal{A}_g(L)$. One instructive example is $\mathcal{M}_3(l)$, where l is odd and ≥ 3. The map $\mathcal{M}_3(l) \to \mathcal{A}_3(l)$ is branched along the hyperelliptic locus. Theorem 15.1 does not apply. However, Nori's Theorem 15.3 does apply—remember, normal functions in weight -1 extend by Theorem 7.1. In this way we realize the normal function associated to λ_3 in Nori's result. Also, by standard arguments, for each n, there is an open subset U of $M_3(l)$ and an unbranched finite cover V of U over which the natural projection $\mathcal{M}_3^n(l) \to \mathcal{M}_3(l)$ has a section. From this one can construct n linearly independent normal sections of the jacobian bundle defined over V. Note that Nori's result does apply to V, whereas Theorem 15.1 does not.

References

1. E. Arbarello and M. Cornalba, *The Picard group of the moduli space of curves*, Topology **26** (1987), 153–171.
2. M. Atiyah, *Riemann surfaces and spin structures*, Ann. Sci. École Norm. Sup. **4** (1971), 47–62.
3. H. Bass, J. Milnor, and J.-P. Serre, *Solution of the congruence subgroup problem for* SL_n $(n \geq 3)$ *and* Sp_{2n} $(n \geq 2)$, Publ. Math. IHES **33** (1967), 59–137.
4. A. Beilinson, *Notes on absolute Hodge cohomology*, Applications of Algebraic K-Theory to Algebraic Geometry and Number Theory (S. J. Bloch et al., eds.), Contemp. Math. **55**, AMS, 1986, part 1, pp. 35–68.
5. ———, *Height pairing between algebraic cycles*, Current Trends in Arithmetical Algebraic Geometry (K. Ribet, ed.), Contemp. Math. **67**, AMS, 1987, pp. 1–24.
6. A. Borel, *Stable real cohomology of arithmetic groups*, Ann. Sci. École Norm. Sup. **7** (1974), 235–272.
7. ———, *Stable real cohomology of arithmetic groups II*, Manifolds and Lie Groups, Papers in Honor of Yozo Matsushima (J. Hano et al., eds.) Prog. in Math. **14**, Birkhäuser, Boston, 1981, pp. 21–55.
8. J.-B. Bost, *Green's currents and height pairing on complex tori*, Duke. Math. J. **61** (1990), 899–912.
9. J. Carlson, *The geometry of the extension class of a mixed Hodge structure*, Algebraic Geometry, Bowdoin, 1985 (S. J. Bloch, ed.), Proc. Sympos. Pure Math. **46**, AMS, 1987, pp. 199–222.
10. G. Ceresa, *C is not algebraically equivalent to C^- in its jacobian*, Ann. of Math. **117** (1983), 285–291.
11. K.-T. Chen, *Extension of C^∞ function algebra and Malcev completion of π_1*, Adv. Math. 23 (1977), 181–210.
12. W. Fulton, *Intersection Theory*, Springer, New York, 1984.
13. W. Fulton, J. Harris, *Representation Theory, a First Course*, Grad. Texts in Math. **129**, Springer, New York, 1991.
14. P. Griffiths, *On the periods of certain rational integrals*, Ann. of Math. **90** (1969), 460–541.
15. F. Guillén,V. Navarro Aznar, P. Pascula-Gainza, F. Puerta, *Hyperrésolutions cubiques et descente cohomologique*, Lecture Notes in Math. **1335**, Springer, Berlin, 1988.
16. R. Hain, *The geometry of the mixed Hodge structure on the fundamental group*, Proc. Sympos. Pure Math. **46**, AMS, 1987, vol. 2, pp. 247–281.
17. ———, *The de Rham homotopy theory of complex algebraic varieties II*, K-Theory **1** (1987), 481–497.
18. ———, *Biextensions and heights associated to curves of odd genus*, Duke. Math. J. **61** (1990), 859–898.
19. ———, *Completions of mapping class groups and the cycle $C - C^-$*, Mapping Class Groups and Moduli Spaces of Riemann Surfaces, (C.-F. Bödigheimer and R. Hain, eds.), Contemp. Math. **150**, AMS, 1993, pp. 75–105.
20. ———, *Classical polylogarithms*, Motives (U. Janssen et al., eds.), Proc. Sympos. Pure Math. **55**, AMS, 1994.
21. J. Harer, *The second homology group of the mapping class group of an orientable surface*, Invent. Math. **72** (1983), 221–239.

22. ———, *Stability of the homology of the mapping class groups of orientable surfaces*, Ann. Math. **121** (1985), 215–249.

23. ———, *The cohomology of the moduli space of curves*, Theory of Moduli, (E. Sernesi, ed.), Lecture Notes in Math. **1337**, Springer, Berlin, 1988, pp. 139–221.

24. ———, *The third homology group of the moduli space of curves*, Duke. Math. J. **63** (1992), 25–55.

25. ———, *The rational Picard group of the moduli spaces of Riemann surfaces with spin structure*, Mapping Class Groups and Moduli Spaces of Riemann Surfaces (C.-F. Bödigheimer and R. Hain, eds.), Contemp. Math. **150**, AMS, 1993, pp. 107–136.

26. ———, *The fourth homology group of the moduli space of curves*, preprint, 1993.

27. B. Harris, *Harmonic volumes*, Acta Math. **150** (1983), 91–123.

28. N. Ivanov, *Complexes of curves and the Teichmüller modular group*, Uspekhi Mat. Nauk **42** (1987), 49–91; English translation: *Russian Math. Surveys* **42** (1987), 55–107.

29. D. Johnson, *An abelian quotient of the mapping class group $\mathcal{I}_g$*, Math. Ann. **249** (1980), 225–242.

30. ———, *The structure of the Torelli group I: A finite set of generators for $\mathcal{I}$*, Ann. of Math. **118** (1983), 423–442.

31. ———, *The structure of the Torelli group—II: A characterization of the group generated by twists on bounding curves*, Topology **24** (1985), 113–126.

32. ———, *The structure of the Torelli group—III: The abelianization of $\mathcal{I}$*, Topology **24** (1985), 127–144.

33. ———, *A survey of the Torelli group*, Low dimensional topology (S. J. Lomonaco, Jr., ed.), Contemp. Math. **20**, AMS, 1983, pp. 165–179.

34. D. Lear, *Extensions of Normal Functions and Asymptotics of the Height Pairing*, Ph.D. Thesis, University of Washington, 1990.

35. S. Mac Lane, *Homology*, Springer, Berlin, 1963.

36. W. Magnus, A. Karras, and D. Solitar, *Combinatorial Group Theory*, Interscience, 1966.

37. G. Mess, *The Torelli groups for genus 2 and 3 surfaces*, MSRI preprint 05608-90, 1990.

38. S. Morita, *The structure of the mapping class group and characteristic classes of surface bundles*, Mapping Class Groups and Moduli Spaces of Riemann Surfaces (C.-F. Bödigheimer and R. Hain, eds.), Contemp. Math. **150**, AMS, 1993.

39. D. Mumford, *Abelian quotients of the Teichmüller modular group*, J. Anal. Math. **18** (1967), 227–244.

40. M. Nori, *Algebraic cycles and Hodge theoretic connectivity*, Invent. Math. **111** (1993), 349–373.

41. M. Pulte, *The fundamental group of a Riemann surface: mixed Hodge structures and algebraic cycles*, Duke. Math. J. **57** (1988), 721–760.

42. M. Ragunathan, *Cohomology of arithmetic subgroups of algebraic groups: I*, Ann. of Math. **86** (1967), 409–424.

43. M. Saito, *Mixed Hodge modules and admissible variations*, C. R. Acad. Sci. Paris, **309** (1989), Série I, 351–356.

44. J.-P. Serre, *Lie Algebras and Lie Groups*, Benjamin, 1965.

45. A. Silverberg, *Mordell–Weil groups of generic abelian varieties*, Invent. Math. **81** (1985), 71–106.

46. P. Sipe, *Some finite quotients of the mapping class group of a surface*, Proc. Amer. Math. Soc. **97** (1986), 515–524.

47. J. Steenbrink and S. Zucker, *Variations of mixed Hodge structure I*, Invent. Math. **80** (1985), 489–542.

48. D. Sullivan, *Infinitesimal computations in topology*, Publ. Math. IHES **47** (1977), 269–331.

Richard M. Hain
Department of Mathematics, Duke University
Durham, NC 27708-0320

E-mail address: hain@math.duke.edu

Complex Algebraic Geometry
MSRI Publications
Volume **28**, 1995

Vector Bundles and Brill–Noether Theory

SHIGERU MUKAI

ABSTRACT. After a quick review of the Picard variety and Brill–Noether theory, we generalize them to holomorphic rank-two vector bundles of canonical determinant over a compact Riemann surface. We propose several problems of Brill–Noether type for such bundles and announce some of our results concerning the Brill–Noether loci and Fano threefolds. For example, the locus of rank-two bundles of canonical determinant with five linearly independent global sections on a non-tetragonal curve of genus 7 is a smooth Fano threefold of genus 7.

As a natural generalization of line bundles, vector bundles have two important roles in algebraic geometry. One is the moduli space. The moduli of vector bundles gives connections among different types of varieties, and sometimes yields new varieties that are difficult to describe by other means. The other is the linear system. In the same way as the classical construction of a map to a projective space, a vector bundle gives rise to a rational map to a Grassmannian if it is generically generated by its global sections. In this article, we shall describe some results for which vector bundles play such roles. They are obtained from an attempt to generalize Brill–Noether theory of special divisors, reviewed in Section 2, to vector bundles. Our main subject is rank-two vector bundles with canonical determinant on a curve C with as many global sections as possible: especially their moduli and the Grassmannian embeddings of C by them (Section 4).

1. Line bundles

Let X be a smooth algebraic variety over the complex number field $\mathbf{C}$. We consider the set of isomorphism classes of line bundles, or invertible sheaves, on X. This set enjoys two good properties, neither of which holds anymore

Based on the author's three talks given at JAMI in 1991, UCLA in 1992 and Durham University in 1993. Supported in part by a Grant under The Monbusho International Science Research Program: 04044081.

for vector bundles of higher rank. One is that it has a natural algebraic structure as a moduli space without any modification. The other is that it becomes a (commutative) group by the tensor product. In fact, the isomorphism classes are parametrized by the first cohomology group $H^1(\mathcal{O}_X^*)$ with coefficient in the (multiplicative) sheaf of nowhere vanishing holomorphic functions. $H^1(\mathcal{O}_X^*)$ endowed with the natural algebraic structure is called the *Picard variety* and denoted by $\operatorname{Pic} X$. Let

$$(1.1) \quad \cdots \longrightarrow H^1(X, \mathbf{Z}) \longrightarrow H^1(\mathcal{O}_X) \longrightarrow H^1(\mathcal{O}_X^*) \stackrel{\delta}{\longrightarrow} H^2(X, \mathbf{Z}) \longrightarrow \cdots$$

be the long exact sequence derived from the exponential exact sequence

$$(1.2) \qquad 0 \longrightarrow \mathbf{Z} \stackrel{2\pi i}{\longrightarrow} \mathcal{O}_X \stackrel{\exp}{\longrightarrow} \mathcal{O}_X^* \longrightarrow 1$$

of sheaves on X. The connecting homomorphism δ associates the first Chern class $c_1(L)$ to each line bundle $[L] \in H^1(\mathcal{O}_X^*)$. For example, if X is a curve, $\delta(L)$ is the degree of L under the natural identification $H^2(X, \mathbf{Z}) \simeq \mathbf{Z}$. By (1.1), the neutral component $\operatorname{Pic}^0 X$ of $\operatorname{Pic} X$ is isomorphic to the quotient group $H^1(\mathcal{O}_X)/H^1(X, \mathbf{Z})$, which is an abelian variety if X is a projective variety.

Let C be a *curve*, or a compact Riemann surface, of genus g. The Riemann–Roch theorem

$$(1.3) \qquad \begin{cases} \chi(L) := h^0(L) - h^1(L) = \deg L + 1 - g, \\ H^1(L) \simeq H^0(K_C L^{-1})^\vee, \end{cases}$$

is most fundamental for its study. The latter isomorphism is functorial in L and referred to as the *Serre duality.* By (1.1), $\operatorname{Pic} C$ is the disjoint union of $\operatorname{Pic}^d C$, for $d \in \mathbf{Z}$, where $\operatorname{Pic}^d C$ is the set of isomorphism classes of line bundles of degree d. By (1.3), the number $h^0(L)$ of linearly independent global sections is constant on $\operatorname{Pic}^d C$ unless $0 \le d \le 2g-2 = \deg K_C$. Conversely, when $0 \le d \le 2g-2$, the number $h^0(L)$ is equal to $\max\{0, d+1-g\}$ on a non-empty Zariski open subset of $\operatorname{Pic}^d C$, but not constant since there exists a *special line bundle*, that is, a line bundle L with $h^0(L)h^1(L) \neq 0$, of degree d. The space $\operatorname{Pic}^d C$ is stratified by $h^0(L)$. For an integer $r \geq \max\{0, d+1-g\}$, we set

$$W_d^r(C) = \{[L] \mid h^0(L) \geq r+1\} \subset \operatorname{Pic}^d C,$$

which is closed in the Zariski topology. The case $(d, r) = (g-1, 0)$ is most important. W_{g-1}^0 is a divisor and usually denoted by Θ. The self-intersection number (Θ^g) is equal to $g!$, i.e., Θ is a principal polarization of $\operatorname{Pic}^{g-1} C$. This principally polarized abelian variety $(\operatorname{Pic}^{g-1} C, \Theta)$ is called the *Jacobian* of C.

Often the isomorphism class of C is recovered from the variety $W_d^r(C)$ of special line bundles. The case of theta divisor Θ is classical:

THEOREM 1.4 (TORELLI). *Two curves are isomorphic to each other if their Jacobians are so.*

We refer to [13] for various approaches to this important result. Let C be a non-hyperelliptic curve of genus 5. Then $W_4^1(C)$ is a curve of genus 11. (If C is trigonal or $W_4^1(C)$ contains a line bundle with $L^2 \simeq K_C$, then $W_4^1(C)$ is singular. But still the theorem holds true.)

Another example is:

THEOREM 1.5. *The Jacobian of C is isomorphic to the Prym variety of*

$$(W_4^1(C), \sigma),$$

where σ is the involution of $\operatorname{Pic}^4 C$ *defined by* $\sigma[L] = [K_C L^{-1}]$.

See [2] for the proof in the case where C is a complete intersection of three quadrics in $\mathbf{P}^5$.

Another feature of special line bundles is their relation with projective embeddings. If a line bundle L is generated by its global sections, we obtain a morphism

$$\Phi_{|L|} : C \longrightarrow \mathbf{P}^* H^0(L),$$

where $\mathbf{P}^* H^0(L)$ is the projectivization of the dual vector space of $H^0(L)$. The most interesting case is K_C, the canonical line bundle, which appears in (1.3). By the Riemann–Roch theorem, K_C is generated by global sections, and $\Phi_{|K|} : C \longrightarrow \mathbf{P}^* H^0(K_C) = \mathbf{P}^{g-1}$ is an embedding unless C is hyperelliptic. The image $C_{2g-2} \subset \mathbf{P}^{g-1}$ of $\Phi_{|K|}$ is called *the canonical model* of C. Here is a classical example:

THEOREM 1.6 (ENRIQUES–PETRI). *The canonical model $C_{2g-2} \subset \mathbf{P}^{g-1}$ is an intersection of quadrics if and only if $W_3^1(C) = W_5^2(C) = \varnothing$.*

We refer to [1] and [6] for further results of this kind. The latter also discusses an interesting use of vector bundles that we do not treat here.

2. Brill–Noether theory

We study $W_d^r(C)$ more closely. First we note that it is not only a subset but a subscheme of $\operatorname{Pic} C$. Take distinct points $P_1, \ldots, P_N$ of C and put $D = \sum_{i=1}^N P_i$. Choose N sufficiently large so that $H^1(L(D))$ vanishes for every $[L] \in \operatorname{Pic}^d C$. The exact sequence

$$0 \longrightarrow L \longrightarrow L(D) \xrightarrow{\text{res}} \bigoplus_{i=1}^{N} L(D)|_{P_i} \longrightarrow 0 \tag{2.1}$$

of sheaves on C induces the exact sequence

$$(2.2)\quad 0 \longrightarrow H^0(L) \longrightarrow H^0(L(D)) \overset{H^0(\mathrm{res})}{\longrightarrow} \bigoplus_{i=1}^{N} H^0(L(D)|_{P_i}) \longrightarrow H^1(L) \longrightarrow 0$$

of vector spaces. There exists a homomorphism $R : \mathcal{E} \longrightarrow \mathcal{F}$ between two vector bundles $\mathcal{E}$ and $\mathcal{F}$ on $\operatorname{Pic}^d C$ whose fibre $R_{[L]}$ at $[L]$ is the above $H^0(\mathrm{res})$ for every $[L] \in \operatorname{Pic}^d C$ (these bundles are the direct images of certain sheaves on $C \times \operatorname{Pic} C$). The difference in rank between $\mathcal{E}$ and $\mathcal{F}$ does not depend on D: we have

$$r(\mathcal{F}) - r(\mathcal{E}) = N - h^0(L(D)) = g - 1 - d$$

by (1.3). The following statement is easy to verify:

LEMMA 2.3. *Let E and F be finite-dimensional vector spaces, let c be a positive integer, and set $W = \{f \in \operatorname{Hom}(E, F) \mid \dim \operatorname{Ker} f \geq c\}$. Then:*
1) *W is a closed subvariety of codimension* $\max\{0,\ c(c+\delta)\}$ *in the affine space* $\operatorname{Hom}(E, F)$, *where $\delta = \dim F - \dim E$, and*
2) *if* $\dim \operatorname{Ker} f = c$, *then W is smooth at the point f and the normal space $N_{W/\mathrm{Hom},f}$ is isomorphic to* $\operatorname{Hom}(\operatorname{Ker} f, \operatorname{Coker} f)$.

Since $W^r_d(C)$ is

$$\{\alpha \in \operatorname{Pic}^d C \mid \dim \operatorname{Ker} R_\alpha \geq r+1\},$$

it is a closed subscheme of $\operatorname{Pic}^d C$ and its codimension is at most $(r+1)(g+r-d)$ by the lemma. It follows that

$$(2.4)\qquad \dim W^r_d(C) \geq g - (r+1)(g+r-d).$$

For a line bundle L on C, we put $\rho(L) = g - h^0(L)h^1(L)$ and call it the *Brill–Noether number.* When $[L] \in W^r_d(C)$ and $h^0(L) = r+1$, this number is equal to the right-hand side of the above inequality. Since the tangent space of $\operatorname{Pic} C$ is isomorphic to $H^1(\mathcal{O}_C)$ by (1.1), the Zariski tangent space of $W^r_d(C)$ at $[L]$ is the kernel of the tangential map $H^1(\mathcal{O}_C) \longrightarrow \operatorname{Hom}(H^0(L), H^1(L))$ by (2.2) and Lemma 2.3(2).

Now we describe the Zariski tangent space more directly. Let

$$\tau_L : H^1(\mathcal{O}_C) \longrightarrow \operatorname{Hom}(H^0(L), H^1(L))$$

be the linear map induced by the cup product $H^1(\mathcal{O}_C) \times H^0(L) \longrightarrow H^1(L)$. By the Serre duality (1.3), the dual of τ_L is the multiplication map

$$(2.5)\qquad H^0(L) \otimes H^0(K_C L^{-1}) \longrightarrow H^0(K_C),$$

called the *Petri map.*

PROPOSITION 2.6. *Assume that* $[L] \in W^r_d(C)$ *and* $h^0(L) = r+1$. *Then:*

1) *the Zariski tangent space of* $W^r_d(C)$ *at* $[L]$ *is isomorphic to* $\operatorname{Ker}\tau_L$,
2) $\dim \operatorname{Ker}\tau_L \geq \dim_{[L]} W^r_d(C) \geq \rho(L)$, *and*
3) *the following three conditions are equivalent:*
   i) $W^r_d(C)$ *is smooth and of dimension* $\rho(L)$ *at* $[L]$,
   ii) τ_L *is surjective, and*
   iii) *the Petri map* (2.5) *is injective,*

PROOF. Let $\alpha = \{a_{ij}\} \in H^1(\mathcal{O}^*_C)$ be the cohomology class corresponding to L, that is, $a_{ij} \in \mathcal{O}_{U_i \cap U_j}$ are the transition functions of L for a suitable open covering $\{U_i\}$ of C. Let ε be the dual number, i.e., $\varepsilon \neq 0$ but $\varepsilon^2 = 0$. A first-order infinitesimal deformation $\tilde{L}$ of L corresponds to a cohomology class $\tilde{\alpha} = \{\tilde{a}_{ij}\} \in H^1(\mathcal{O}_C[\varepsilon]^*)$ whose reduction modulo ε is α. Also, $\tilde{\alpha}$ is of the form $\{a_{ij}(1 + b_{ij}\varepsilon)\}$ for $\beta = \{b_{ij}\} \in H^1(\mathcal{O}_C)$. Let $h \in H^0(L)$ be a global section of L; then h is a collection $\{h_i\}$ of $h_i \in \mathcal{O}_{U_i}$ such that $h_i = a_{ij}h_j$. The differences $b_{ij}h_j\varepsilon$ of h_i and $\tilde{a}_{ij}h_j$ form a one-cocycle whose cohomology class is the cup product $(\beta^{\cup} h)\varepsilon$. Hence h extends to a global section of $\tilde{L}$ if and only if $\beta^{\cup} h = 0$ in $H^1(L)$. Therefore, all global sections of L extend if and only if the cup product map ${}^{\cup}\beta : H^0(L) \longrightarrow H^1(L)$ is zero, which shows (1). Part (2) follows from (1) and (2.4). Part (3) is straightforward from (2). □

Let ρ be the right-hand side of (2.4). We refer to [1] for the following important results:

THEOREM 2.7 (Kempf Kleiman–Laksov; Fulton–Lazarsfeld).

(*Existence*) $W^r_d(C) \neq \varnothing$ *if* $\rho \geq 0$.

(*Connectedness*) $W^r_d(C)$ *is connected if* $\rho > 0$.

Let $\mathcal{M}_g$ be the moduli space of curves of genus g.

THEOREM 2.8 (Gieseker [5], Lazarsfeld [7]). *If* $[C] \in \mathcal{M}_g$ *is general, the Petri map* (2.5) *is injective for every* (*special*) *line bundle* L *on* C.

In particular, $W^r_d(C)$ is of dimension ρ if ρ is nonnegative, and empty otherwise. Thus the estimate (2.4) is best possible for the generic curve. When $\rho = 0$, the number of $W^r_d(C)$ is finite and was first computed by Castelnuovo. Let $G(a,\, a+b) \subset \mathbf{P}_* \bigwedge^a \mathbf{C}^{a+b}$ be the Plücker embedding of the Grassmannian of a-dimensional subspaces of $\mathbf{C}^{a+b}$. The following result is interesting (cf. (4.15)):

THEOREM 2.9. *If* $[C] \in \mathcal{M}_g$ *is general and* $\rho = 0$, *the number of* $W^r_d(C)$ *is equal to the degree of the* g*-dimensional Grassmannian*

$$G(a, a+b) \subset \mathbf{P}_* \bigwedge^a \mathbf{C}^{a+b},$$

where $a = r + 1$ *and* $b = g + r - d$.

In fact, both $\#W_d^r(C)$ and $\deg G(a, a+b)$ are equal to

$$g! \prod_{1 \le i \le a < j \le a+b} (j-i)^{-1}.$$

3. Vector bundles on a curve

Let X be a smooth complete algebraic variety (over $\mathbf{C}$). By $\mathcal{O}_X$, we mean either the sheaf of holomorphic functions in the usual topology or the sheaf of regular functions in the Zariski topology. In the study of vector bundles, this is allowed by virtue of the GAGA principle, which says that the two categories of analytic and algebraic coherent sheaves on X are equivalent to each other. By a *vector bundle* E on X we mean a locally free $\mathcal{O}_X$-module. There exists an open covering $\{U_i\}$ of X such that $E|_{U_i} \simeq \mathcal{O}_{U_i}^{\oplus r}$ for every i. This positive integer r is called the *rank* of E. The highest exterior product $\bigwedge^r E$ is a line bundle on X, denoted by $\det E$.

Assume that E is generated by global sections, that is, the evaluation homomorphism

$$\mathrm{ev}_E : H^0(E) \otimes \mathcal{O}_X \longrightarrow E$$

is surjective. Then every fibre E_x of E is an r-dimensional quotient space of $H^0(E)$. Hence we obtain a map $\Phi_{|E|} : X \longrightarrow G(H^0(E), r)$ to the Grassmannian of r-dimensional *quotient* spaces of $H^0(E)$. This map is holomorphic since E is so. The exterior product

$$\bigwedge^r \mathrm{ev}_E : \bigwedge^r H^0(E) \otimes \mathcal{O}_X \longrightarrow \bigwedge^r E$$

of ev_E induces a linear map

$$\bigwedge^r H^0(E) \longrightarrow H^0(\bigwedge^r E), \tag{3.1}$$

which we denote by λ_E. The exterior product $\bigwedge^r E_x$ of fibres E_x are quotient spaces of both $H^0(\bigwedge^r E)$ and $\bigwedge^r H^0(E)$. The former determines a point in the projective space $\mathbf{P}^* H^0(\bigwedge^r E)$ and the latter the Plücker coordinate of $[E_x] \in G(H^0(E), r)$. Hence we obtain the following commutative diagram:

$$\begin{array}{ccc} X & \xrightarrow{\Phi_{|E|}} & G(H^0(E), r) \\ \Phi_{|\wedge^r E|} \downarrow & & \cap \text{ Plücker embedding} \\ \mathbf{P}^* H^0(\bigwedge^r E) & \overset{\mathbf{P}^*\lambda_E}{\cdots\rightarrow} & \mathbf{P}^* \bigwedge^r H^0(E). \end{array} \tag{3.2}$$

Thus λ_E connects the projective and Grassmannian embeddings. This map is important in other considerations, too. See (3.5), (4.7) and (4.15).

Let $A_{ij} \in \mathrm{GL}(r, \mathcal{O}_{U_i \cap U_j})$ be the transition matrix function of a vector bundle E, that is, the matrix expression of the composite of the two isomorphisms $\mathcal{O}_{U_j}^{\oplus r} \simeq E|_{U_j}$ and $E|_{U_i} \simeq \mathcal{O}_{U_i}^{\oplus r}$ over $U_i \cap U_j$. The collection $\{A_{ij}\}$ of all such (matrix) functions is a one-cocycle with coefficients in the sheaf $\mathrm{GL}(r, \mathcal{O}_X)$ of noncommutative groups. Hence the rank-r vector bundles on X are parametrized by the first cohomology set $H^1(\mathrm{GL}(r, \mathcal{O}_X))$. The determinant homomorphism $\det : \mathrm{GL}(r, \mathcal{O}_X) \longrightarrow \mathcal{O}_X^*$ induces the map

$$H^1(\det) : H^1(\mathrm{GL}(r, \mathcal{O}_X)) \longrightarrow H^1(\mathcal{O}_X^*) = \operatorname{Pic} X,$$

whose fibre at $[L] \in \operatorname{Pic} X$ is denoted by $B_X(r, L)$.

Moduli Problem. a) Give a natural algebraic structure to a suitable *open* subset of $B_X(r, L)$ and construct its *geometric* compactification.

b) What properties of (X, L) are inherited by the moduli space constructed in (a)?

The fibre $B_X(r, L)$ does not have a nice description such as $\operatorname{Pic} X$ in (1.1). But its Zariski tangent space as a functor is easy to identify. Let E be a vector bundle and $\{A_{ij}\}$ the one-cocycle of transition functions. A one-cochain $\{A_{ij}(I_r + B_{ij}\varepsilon)\}$ with values in $\mathrm{GL}(r, \mathcal{O}_X[\varepsilon])$ is a cocycle if and only if

$$A_{jk}^{-1} B_{ij} A_{jk} + B_{jk} = B_{ik}$$

holds on $U_i \cap U_j \cap U_k$ for every i, j and k, where ε is the dual number. This is the same as saying that $\{B_{ij}\}$ is a one-cocycle with values in the sheaf $\mathcal{E}nd\, E \simeq E^\vee \otimes E$ of (local) endomorphisms of E. By this correspondence, the first-order infinitesimal deformations of E are parametrized by the cohomology group $H^1(\mathcal{E}nd\, E)$. Since

$$\det(A_{ij}(I_r + B_{ij}\varepsilon)) = (\det A_{ij})(I_r + \operatorname{tr} B_{ij}\varepsilon),$$

the Zariski tangent space of $B_X(r, L)$ is isomorphic to the kernel of

$$H^1(\mathrm{tr}) : H^1(\mathcal{E}nd\, E) \longrightarrow H^1(\mathcal{O}_X),$$

which is the tangential map of $H^1(\det)$. Let $\mathrm{sl}(E)$ be the sheaf of traceless endomorphisms of E. Then $\mathcal{E}nd\, E$ is the direct sum of two vector bundles $\mathrm{sl}(E)$ and $\mathcal{O}_X$. Therefore, the Zariski tangent space is isomorphic to $H^1(\mathrm{sl}(E))$.

Let C be a curve of genus g. The Riemann–Roch theorem (1.3) is generalized to the formula

$$(3.3) \qquad \begin{cases} \chi(E) := h^0(E) - h^1(E) = \deg E + r(1-g) \\ H^1(E) \simeq H^0(K_C E^\vee)^\vee \quad \text{(Serre duality)}, \end{cases}$$

where $\deg E$ is the degree of $\det E$. For simplicity, we restrict ourselves to the case $r = 2$. The answer to part (a) of the moduli problem is the notion of stability:

DEFINITION 3.4. (Mumford [11]) A rank-two vector bundle E on C is *stable* if $\deg \xi < \frac{1}{2} \deg E$ for every line subbundle ξ of E. It is *semi-stable* if $\deg \xi \leq \frac{1}{2} \deg E$ for every ξ.

Let L be a line bundle on C and E a member of $B_C(2, L)$. Fix an ample line bundle $\mathcal{O}_C(1)$ on C. Then $E(n)$ belongs to $B_C(2, L(2n))$ and we obtain the linear map

$$\lambda_{E(n)} : \bigwedge^2 H^0(E(n)) \longrightarrow H^0(L(2n)) \tag{3.5}$$

as in (3.1). The above condition $\deg \xi < \frac{1}{2} \deg E$ is equivalent to the asymptotic stability of the linear map $\lambda_{E(n)}$ with respect to the action of the special linear group $\mathrm{SL}(H^0(E(n)))$ [4]. By the geometric invariant theory, the (coarse) moduli space $M_C(2, L)$ of stable two-bundles with determinant L exists as a quasi-projective algebraic variety. Moreover, it becomes a projective algebraic variety $\overline{M}_C(2, L)$ by adding certain equivalence classes of semi-stable two-bundles.

Every line bundle M induces an isomorphism $M_C(2, L) \simeq M_C(2, LM^2)$ taking E to $E \otimes M$. Hence there are only two isomorphism classes of the moduli space of two-bundles: $M_C(2, \mathrm{odd})$ and $M_C(2, \mathrm{even})$. Both are smooth and of dimension $3g - 3 = \dim H^1(\mathrm{sl}(E))$. Since every semi-stable two-bundle of odd degree is stable, $M_C(2, \mathrm{odd})$ is a projective variety. Among many known global properties of $M_C(2, \mathrm{odd})$, we state two. One is

THEOREM 3.6 (Ramanan [19]). *$M_C(2, \mathrm{odd})$ is a Fano manifold of index two.*

A smooth projective variety X is called a *Fano manifold* if the anti-canonical line bundle $K_X^{-1} = \det T_X$ is ample. The largest integer that divides $c_1(X)$ in $H^2(X, \mathbf{Z})$ is called the index. In the case of $M_C(2, \mathrm{odd})$, the Picard group is free cyclic and the anti-canonical line bundle is the square of the positive generator.

EXAMPLE. (Desale and Ramanan [21], Newstead [17]) Let C be a curve of genus two defined by the equation $y^2 = (x - \lambda_1)(x - \lambda_2) \cdots (x - \lambda_6)$. Then the moduli space $M_C(2, \mathrm{odd})$ is the complete intersection

$$\sum_{i=1}^{6} x_i^2 = \sum_{i=1}^{6} \lambda_i x_i^2 = 0$$

in $\mathbf{P}^5$. The anti-canonical line bundle is the square of the restriction of tautological line bundle.

The other result we cite is a Torelli-type theorem:

THEOREM 3.7 (Mumford and Newstead [15]). *The intermediate Jacobian*

$$H^2(\Omega^1_M)/H^3(M,\mathbf{Z})$$

of the moduli space $M_C(2,\mathrm{odd})$ *is isomorphic to the Jacobian of* C *as polarized abelian variety.*

4. Toward a Brill–Noether type theory for two-bundles

$M_C(2,\mathcal{O}_C)$ and $M_C(2,K_C)$ are two natural representatives of $M_C(2,\mathrm{even})$. The following is fundamental for the former:

THEOREM 4.1 (Narasimhan and Seshadri [16], Donaldson [3]). *There is a natural bijection between the two sets*

1) $M_C(2,\mathcal{O}_C)$*, the set of isomorphism classes of stable two-bundles with trivial determinant, and*
2) $\mathrm{Hom}^{\mathrm{irr}}(\pi_1(C),\mathrm{SU}(2))/\mathrm{SU}(2)$*, the set of conjugacy classes of irreducible two-dimensional special unitary representations of the fundamental group of* C.

We take $M_C(2,K)$ to develop a Brill–Noether type theory. See [20] and [18] for another direction of development. The moduli space $M_C(2,K)$ of stable two-bundles with canonical determinant is stratified by the number $h^0(E)$ of linearly independent global sections. By analogy with W^r_d, we set

$$M_C(2,K,n)=\{[E]\mid h^0(E)\geq n+2\}\subset M_C(2,K).$$

We denote by $\overline{M}_C(2,K,n)$ the union of $M_C(2,K,n)$ and the set of isomorphism classes of semi-stable vector bundles $[\xi\oplus K_C\xi^{-1}]\in\overline{M}_C(2,K)$ with $[\xi]\in W^{n/2}_{g-1}(C)$.

By the same argument as in Section 2, $M_C(2,K,n)$ is a closed subscheme of $M_C(2,K)$. (The universal family does not exist on $C\times M_C(2,K)$, but this does not cause a problem for the study of such local properties of the moduli.) A similar consideration gives the estimate $\dim M_C(2,K,n)\geq 3g-3-(n+2)^2$. But this estimate is not sharp. The proper one is

THEOREM 4.2. $\dim M_C(2,K,n)\geq 3g-3-\frac{1}{2}(n+2)(n+3)$.

Before giving the proof, we recall some notions of symplectic geometry. Let W be a 2ν-dimensional vector space with a non-degenerate skew-symmetric bilinear form $\langle\,,\rangle : W\times W\longrightarrow\mathbf{C}$. A ν-dimensional subspace V of W is *a Lagrangian* of W if the bilinear form $\langle\,,\rangle$ is identically zero on $V\times V$. We denote the set of Lagrangians by $\mathcal{L}(W)$, which is a subset of the ν^2-dimensional Grassmannian $G(\nu,W)$. Fix a Lagrangian U_∞ and set $Z=\{[U]\mid U\cap U_\infty=0\}$ in $G(\nu,W)$. When $[U_0]\in Z$ is fixed, Z is a ν^2-dimensional affine space by the bijection

$$\mathrm{Hom}(U_0,U_\infty)\ni f\mapsto\Gamma_f\in Z, \tag{4.3}$$

where $\Gamma_f \subset U_0 \times U_\infty$ is the graph of f. Assume that U_0 is also a Lagrangian. Then U_0 and U_∞ are dual under the pairing $\langle\, ,\rangle$. Γ_f is a Lagrangian of W if and only if $f \in \operatorname{Hom}(U_0, U_\infty) \simeq U_\infty \otimes U_\infty$ is symmetric. Hence $\mathcal{L}(W)$ is a smooth subvariety of dimension $\frac{1}{2}\nu(\nu+1)$ in the Grassmannian $G(\nu, W)$.

PROPOSITION 4.4. *Fix a Lagrangian $[V_0] \in \mathcal{L}(W)$ and let c be a positive integer. Then the Schubert subvariety*

$$\mathcal{L}(W)_c = \{[V] \in \mathcal{L}(W) \mid \dim V \cap V_0 \geq c\}$$

is of codimension $\frac{1}{2}c(c+1)$ in $\mathcal{L}(W)$.

PROOF. For $[V] \in \mathcal{L}(W)$, choose a Lagrangian V_∞ so that $V \cap V_\infty = V_0 \cap V_\infty = 0$. Then V corresponds to a symmetric matrix of size ν via (4.3) and $\dim V \cap V_0$ is equal to the co-rank of the symmetric matrix. Hence we have our assertion. □

PROOF OF THEOREM 4.2. Let E be a two-bundle with canonical determinant and D an effective divisor on C. Since E is self-Serre adjoint, that is, $E^\vee K_C \simeq E$, the two vector spaces $H^0(E)$ and $H^1(E)$ are dual by (3.3). Similarly, $H^0(E(-D))$ and $H^1(E(D))$ are dual. Hence, by the Riemann–Roch theorem (3.3), we have

$$h^0(E(D)) - h^0(E(-D)) = \chi(E(D)) = 2N, \tag{4.5}$$

where $N = \deg D$. Now we denote the quotient sheaf $E(D)/E(-D)$ by A, which is supported by a finite set and has length $4N$. We consider the composite of the pairing $A \times A \longrightarrow K_C(2D)/K_C$, induced by $\bigwedge^2 E \simeq K_C$, and the residue map $r : K_C(2D)/K_C \longrightarrow \mathbf{C}$ given by

$$r(\omega) = \sum_{P \in \operatorname{Supp} D} \operatorname{Res}_P \omega.$$

This induces a non-degenerate skew-symmetric pairing $\langle\, ,\rangle$ on the vector space $H^0(A)$ of dimension $4N$. Since r is identically zero on the image of $H^0(K_C(2D))$ (by the Residue Theorem), so is $\langle\, ,\rangle$ on the image V of $H^0(E(D)) \longrightarrow H^0(A)$. By the exact sequence

$$0 \longrightarrow E(-D) \longrightarrow E(D) \longrightarrow A \longrightarrow 0,$$

the image V is isomorphic to the quotient space $H^0(E(D))/H^0(E(-D))$. Hence V is a Lagrangian of the symplectic vector space $H^0(A)$ by (4.5). It is obvious that $V_0 = H^0(E/E(-D))$ is also a Lagrangian. Now we choose D so that $H^0(E(-D)) = 0$ for every $[E] \in M_C(2, K)$. Then $H^0(E)$ is the intersection of two Lagrangians $V = H^0(E(D))$ and V_0 of $H^0(A)$. Hence we have our inequality by the above proposition. □

REMARK 4.6. This proof works also for a vector bundle E of even rank with a non-degenerate skew-symmetric pairing $E \times E \longrightarrow K_C$. The same trick was used in [12] to show the parity preservation of $h^0(E)$ when E has a non-degenerate quadratic form $q : E \longrightarrow K_C$.

Let E be a rank-two vector bundle on C. For $f \in \operatorname{Hom}(H^0(E), H^1(E))$, let $T(f)$ be the linear map

$$\bigwedge^2 H^0(E) \ni h_1 \wedge h_2 \quad \mapsto \quad h_1{}^{\cup} f(h_2) - h_2{}^{\cup} f(h_1) \in H^1\Bigl(\bigwedge^2 E\Bigr),$$

where ${}^{\cup} : H^0(E) \times H^1(E) \longrightarrow H^1(\bigwedge^2 E)$ is the cup product. It is easy to check that the following diagram is commutative:

$$\begin{array}{ccc} H^1(\mathcal{E}nd\, E) & \longrightarrow & \operatorname{Hom}(H^0(E), H^1(E)) \\ {\scriptstyle H^1(\mathrm{tr})} \downarrow & & \downarrow T \\ H^1(\mathcal{O}_C) & \longrightarrow & \operatorname{Hom}(\bigwedge^2 H^0(E), H^1(\bigwedge^2 E)), \end{array} \tag{4.7}$$

where the lower horizontal linear map is the composite of

$$\tau_{\det E} : H^1(\mathcal{O}_C) \longrightarrow \operatorname{Hom}\Bigl(H^0\Bigl(\bigwedge^2 E\Bigr),\, H^1\Bigl(\bigwedge^2 E\Bigr)\Bigr),$$

defined in Section 2, and $\operatorname{Hom}\bigl(\lambda_E,\, H^1(\bigwedge^2 E)\bigr)$.

THEOREM 4.8. *Assume that* $[E] \in M_C(2, K, n)$ *and* $h^0(E) = n + 2$. *Then:*

(1) *the Zariski cotangent space of* $M_C(2, K, n)$ *at* $[E]$ *is isomorphic to the cokernel of* $S^2 H^0(E) \longrightarrow H^0(S^2 E)$,
(2) $\dim \operatorname{Coker}[S^2 H^0(E) \longrightarrow H^0(S^2 E)] \geq \dim_{[E]} M_C(2, K, n) \geq \sigma(E)$, *and*
(3) $M_C(2, K, n)$ *is smooth and of dimension* $\sigma(E)$ *at* $[E]$ *if and only if the map* $S^2 H^0(E) \longrightarrow H^0(S^2 E)$ *is injective.*

PROOF. As we saw in Section 3, the tangent space of $M_C(2, K)$ at $[E]$ is the kernel $H^1(\mathrm{sl}(E))$ of the trace map $H^1(\mathcal{E}nd\, E) \longrightarrow H^1(\mathcal{O}_C)$. Let $\tilde{E}$ be a first-order infinitesimal deformation of E corresponding to $B = \{B_{ij}\} \in H^1(\mathcal{E}nd\, E)$. By the same argument as in the proof of Proposition 2.6, a global section $h \in H^0(E)$ extends that of $\tilde{E}$ if and only if the cup product $h^{\cup} B \in H^1(E)$ vanishes. Hence, all global sections of E extend if and only if the cup product map ${}^{\cup}B : H^0(E) \longrightarrow H^1(E)$ is zero. Therefore, the Zariski tangent space of $M_C(2, K, n)$ is the kernel of $H^1(\mathrm{sl}(E)) \longrightarrow \operatorname{Ker} T \subset \operatorname{Hom}(H^0(E), H^1(E))$. Since $\bigwedge^2 E \simeq K_C$, the dual of (4.7) is reduced to the commutative diagram

$$\begin{array}{ccc} H^0(E \otimes E) & \longleftarrow & H^0(E) \otimes H^0(E) \\ \uparrow & & \uparrow \\ H^0(\bigwedge^2 E) & \overset{\lambda_E}{\longleftarrow} & \bigwedge^2 H^0(E), \end{array}$$

by Serre duality (3.3), which shows (1). Part (2) follows from (1) and Theorem 4.2. Part (3) is straightforward from (1) and (2). □

REMARK 4.9. The obstructions for $M_C(2, K, n)$ to be smooth at $[E]$ lie in the cokernel of $H^0(S^2E)^\vee \longrightarrow S^2H^0(E)^\vee$. This fact gives another proof of Theorem 4.2 and 4.8.

Let σ be the right-hand side of Theorem 4.2. Theorem 2.7, 2.8 and 2.9 lead us to the following problems:

PROBLEM 4.10.

(*Existence*) Is $\overline{M}_C(2, K, n)$ non-empty when $\sigma \geq 0$?

(*Connectedness*) Is $\overline{M}_C(2, K, n)$ connected when $\sigma > 0$?

PROBLEM 4.11. Assume that $[C] \in \mathcal{M}_g$ is general.

1) Is $S^2H^0(E) \longrightarrow H^0(S^2E)$ injective for every $E \in M_C(2, K)$?
2) Is $M_C(2, K, n)$ of dimension σ when $\sigma \geq 0$?
3) Compute the number of $M_C(2, K, n)$ when $\sigma = 0$. More generally, describe the cohomology class of $\overline{M}_C(2, K, n)$ in $H^*(\overline{M}_C(2, K), \mathbf{Z})$ when $\sigma > 0$.

Another direction is

PROBLEM 4.12. Study the Grassmannian map associated with a member of $M_C(2, K, n)$, and its relation with the canonical model $C_{2g-2} \subset \mathbf{P}^{g-1}$.

We give some sample results in these directions. They are closely related to our classification of Fano threefolds via vector bundles [8]. We first consider the three cases $(g, n + 2) = (7, 5)$, $(9, 6)$ and $(11, 7)$. The Brill–Noether number σ is equal to 3, 3 and 2, respectively.

THEOREM 4.13. *Let C be a curve of genus 7 with $W^1_4(C) = \varnothing$. Then:*

1) *$M_C(2, K, 3)$ is smooth, complete and of dimension 3;*
2) *$M_C(2, K, 3)$ is a Fano threefold of genus 7, i.e., $(-K_M)^3 = 12$, and with Picard number one; and*
3) *the intermediate Jacobian $H^2(\Omega^1_M)/H^3(M, \mathbf{Z})$ of $M_C(2, K, 3)$ is isomorphic to the Jacobian of C as polarized abelian variety.*

Conversely, every smooth Fano threefold of genus 7 with Picard number one is obtained in this manner from a non-tetragonal curve of genus 7.

Similarly, $\overline{M}_C(2, K, 4)$ is a quartic threefold in $\mathbf{P}^4$ with 21 singular points at the boundary if C is a general curve of genus 9, and $M_C(2, K, 5)$ is a (polarized) K3 surface of genus 11 if C is a general curve of genus 11.

In the case $(g, n + 2) = (8, 6)$, the number σ is equal to zero.

THEOREM 4.14 (Mukai [9], [10]). *If C is a curve of genus 8 with $W^2_7(C) = \varnothing$, then $M_C(2, K, 4)$ consists of the unique isomorphism class of stable two-bundles E. The linear map λ_E in (3.1) is surjective and the following diagram, essentially (3.2), is Cartesian:*

$$\begin{array}{ccc} C & \xrightarrow{\Phi_{|E|}} & G(H^0(E), 2) \\ \text{canonical embedding } \cap & & \cap \text{ Plücker embedding} \\ \mathbf{P}^* H^0(K_C) & \xrightarrow{\mathbf{P}^*\lambda_E} & \mathbf{P}^* \bigwedge^2 H^0(E). \end{array}$$

In particular, C is a complete linear section of the 8-dimensional Grassmannian, that is,

$$[C \subset \mathbf{P}^7] = [G(6,2) \subset \mathbf{P}^{14}] \cap H_1 \cap \cdots \cap H_7$$

for seven hyperplanes $H_1, \ldots, H_7$.

Let C and E be as in the theorem and consider the intersection of $G(2, H^0(E))$ and $\mathbf{P}_* \operatorname{Ker} \lambda_E$, where $G(2, H^0(E))$ is the Grassmannian of two-dimensional subspaces of $H^0(E)$ embedded into $\mathbf{P}_* \bigwedge^2 H^0(E)$ by the Plücker coordinates. If a subspace $[U] \in G(2, H^0(E))$ belongs to $\mathbf{P}_* \operatorname{Ker} \lambda_E$, the evaluation homomorphism $\mathrm{ev}^U : U \otimes \mathcal{O}_C \longrightarrow E$ is not injective and its kernel is a line bundle. Moreover, the inverse of $\operatorname{Ker} \mathrm{ev}^U$ belongs to $W^1_5(C)$, and if C is general, the map

$$(4.15) \qquad G(2, H^0(E)) \cap \mathbf{P}_* \operatorname{Ker} \lambda_E \longrightarrow W^1_5(C), \quad [U] \mapsto (\operatorname{Ker} \mathrm{ev}^U)^{-1}$$

is an isomorphism between two reduced zero-dimensional schemes, which shows Theorem 2.9 in the case $(a, b) = (2, 4)$. This idea leads us to a computation-free proof of Theorem 2.9, which we will discuss elsewhere.

REFERENCES

1. E. Arbarello, M. Cornalba, P. A. Griffiths, and J. Harris, *Geometry of Algebraic Curves I*, Springer, New York, 1985.
2. A. Beauville, *Variétés de Prym et jacobiennes intermédiaires*, Ann. Sci. École Norm. Sup. (4), **10** (1977), 309–391.
3. S. K. Donaldson, *A new proof of a theorem of Narasimhan and Seshadri*, J. Diff. Geom. **18** (1983), 269–2277.
4. D. Gieseker, *On the moduli of vector bundles on an algebraic surface*, Ann. of Math. **106** (1977), 45–60.
5. ———, *Stable curves and special divisors: Petri's conjecture*, Invent. Math., **66** (1982), 251–275.
6. R. Lazarsfeld, *A sampling of vector bundle techniques in the study of linear system*, Lectures on Riemann surfaces, Trieste, 1987 (M. Cornalba et al., eds.), World Scientific, Singapore, 1989, pp. 500–559.
7. ———, *Brill–Noether–Petri without degenerations*, J. Diff. Geom. **23** (1986), 299–307.
8. S. Mukai, *Fano 3-folds*, Complex Projective Geometry (G. Ellingsrud, ed.), London Math. Soc. Lecture Note Series **179**, Cambridge University Press, 1992, pp. 255–263.

9. ———, *Curves and symmetric spaces*, Proc. Japan Acad. **68** (1992), 7–10.

10. ———, *Curves and Grassmannians*, Algebraic Geometry and Related Topics, Inchon, Korea, 1992, International Press, Boston, 1993, pp. 19–40.

11. D. Mumford, *Projective invariants of projective structures and applications*, Internat. Cong. Math. Stockholm, 1962, pp. 526–530.

12. ———, *Theta characteristic of an algebraic curve*, Ann. Sci. École Norm. Sup. (4), **4** (1971), 181–192.

13. ———, *Curves and their Jacobians*, The University of Michigan Press, 1975.

14. ——— and J. Fogarty, *Geometric Invariant Theory*, second edition, Springer, 1982.

15. ——— and P. E. Newstead, *Periods of a moduli space of bundles on a curve*, Amer. J. Math. **90** (1968), 1201–1208.

16. M. S. Narasimhan and Seshadri C. S., *Stable and unitary vector bundles on a compact Riemann surface*, Ann. of Math. **82** (1965), 540–564.

17. P. E. Newstead, *Stable bundles of rank 2 and odd degree over a curve of genus 2*, Topology **7** (1968), 205–215.

18. ———, *Brill–Noether Problems List Update*, University of Liverpool, 1992.

19. S. Ramanan, *The moduli space of vector bundles over an algebraic curve*, Math. Ann. **200** (1973), 69–84.

20. M. Teixidor, *Brill–Noether theory for stable vector bundles*, Duke Math. J., **62** (1991), 385–400.

21. I. V. Desale and S. Ramanan, *Classification of vector bundles of rank 2 on nyperelliptic curves*, Invent. Math., **38** (1976), 161–185.

ADDED IN PROOF. Our determinantal description of $M_C(2, K, n)$ is similar to that of the loci of special divisors in Prym varieties, for which Problems 4.10 and 4.11 have already been solved. See:

22. G. E. Welters, *A theorem of Gieseker–Petri type for Prym varieties*, Ann. Sci. École Norm. Sup. (4), **18** (1985), 671–683.

23. A. Bertram, *An existence theorem for Prym special divisors*, Invent. Math. **90** (1987), 669–671.

24. C. de Concini and P. Pragacz, *On the class of Brill–Noether loci for Prym varieties*, preprint.

SHIGERU MUKAI
DEPARTMENT OF MATHEMATICS, SCHOOL OF SCIENCE
NAGOYA UNIVERSITY
464-01 FURŌ-CHŌ, CHIKUSA-KU
NAGOYA, JAPAN

E-mail address: mukai@math.nagoya-u.ac.jp

For EU product safety concerns, contact us at Calle de José Abascal, 56–1°, 28003 Madrid, Spain or eugpsr@cambridge.org.

www.ingramcontent.com/pod-product-compliance
Ingram Content Group UK Ltd.
Pitfield, Milton Keynes, MK11 3LW, UK
UKHW022133080726
473066UK00009B/530

* 9 7 8 1 1 0 7 4 0 3 8 4 0 *